Scientists in the laboratory often fail to take advantage of the commercial exploitation of their research. This is frequently because they simply do not know what to do. *Technology transfer* is a careful account of how to start the process of commercialisation of technology, and describes in detail the difficulties and the amount of time needed to carry the process through to a successful conclusion. This book provides a much needed step by step guide to the commercialisation of research. It addresses three major themes: how to protect your intellectual property; how to develop it commercially via licensing and business 'start up'; and how to finance and manage your new company. This book is essential reading for any research scientist whose work has commercial applications.

Dr Neil Sullivan has completed a BSc in Biochemistry with Chemistry at the University of Southampton, a Masters of Business Administration (MBA) from Imperial College and a PhD in molecular biology from the University of Edinburgh. He has worked in academic environments at Cold Spring Harbor Laboratories in New York and for the Imperial Cancer Research Fund, London. In addition, he has had practical business experience in both the pharmaceutical and biotechnology industries, together with a recent business development and technology transfer role in a major UK University. He believes that efficient and effective technology transfer is a vital component of the path to long-term success of high technology industries, being a cornerstone of national competitiveness. This book will help guide *you* along this route.

SMG 10/08

Goal: Lead a biotech start up based on my own discoveries.

Technology transfer

What I've Accomplished:

Step 1: Drop out of LHS May 2005 SMG 10/08

Step 2: UMBC ChemE; 3 yrs May 2008 SMG 10/08

Step 3: Cambridge MPhil ACE SMG 10/08

Step 4:

Strategy: SMG 22/10/2008

1). Competitive Advantage: Leave LHS early

2). Lifescience, Engineering, & Technical Skills Training (UMBC ChemE)

3). Business Sense, Management Skills Training (Cambridge ACE)

4). Appreciate all sides of the Scientific and Business Environment (Cambridge ACE).

5). Medical Training & Research Training (MD/PhD)

6). Research Lab → Idea → Innovation → Startup.

7). Sustainable & Profitable Management of Startup

Technology transfer

Making the most of your intellectual property

NEIL F. SULLIVAN

CAMBRIDGE
UNIVERSITY PRESS

CAMBRIDGE UNIVERSITY PRESS
Cambridge, New York, Melbourne, Madrid, Cape Town, Singapore, São Paulo

Cambridge University Press
The Edinburgh Building, Cambridge CB2 2RU, UK

Published in the United States of America by Cambridge University Press, New York

www.cambridge.org
Information on this title: www.cambridge.org/9780521460668

© Cambridge University Press 1995

This publication is in copyright. Subject to statutory exception
and to the provisions of relevant collective licensing agreements,
no reproduction of any part may take place without
the written permission of Cambridge University Press.

First published 1995

A catalogue record for this publication is available from the British Library

Library of Congress Cataloguing in Publication data

Sullivan, Neil F., 1959–
Technology transfer : making the most of your intellectual property / Neil F. Sullivan.
p. cm.
ISBN 0 521 46066 2 (hc). – ISBN 0 521 46616 4 (pb)
1. Technology transfer – Economic aspects. 2. Intellectual property. 3. Research – Economic aspects. 4. Science – Economic aspects. 5. Business planning. I. Title.
HC79.T4S85 1995
658.5′75 – dc20 95–20343 CIP

ISBN-13 978-0-521-46066-8 hardback
ISBN-10 0-521-46066-2 hardback

ISBN-13 978-0-521-46616-5 paperback
ISBN-10 0-521-46616-4 paperback

Transferred to digital printing 2006

Contents

Prologue

The first step in the commercialisation of your technology is one of self audit. Deciding what your 'endpoint' will be is crucial to the determination of the direction in which you move and the methods you take to get there. Any commercialisation will require time and effort and in the initial stages will mostly be your own. The apparently unproductive and ephemeral activities necessary to bring your technology to market will inevitably take you away from the bench and your beloved research. It is thus easy to resent the time spent in commercial exploitation, particularly if the latter is going neither well nor quickly nor if you feel out of your depth in dealing with people of a different mind set. This would perhaps be a mistake, since great satisfaction can be derived from devolving the benefits of your research. Indeed, some would see the effective transfer of research into society as a necessary and integral part of the research process.

This book is aimed at helping researchers to understand the commercial potential of their technology. It will help to bridge the communication gap and familiarise the reader with the plethora of functions, skills and processes that commercialisation requires. It will not be a substitute for the professional advice of patent agents, solicitors, marketing executives, etc., but will go some way to helping you to understand what they require. In addition, the option of 'going it alone' to set up your own business will be evaluated. Many successful companies have been set up on the basis of research and ideas much like your own. Could you raise the finance, structure the business, produce a product, market it, sell it? This book will help you to decide whether **you** could become a 'captain of the biotechnology industry'.

The intention of this book is to point the way towards success. No book can *guarantee* success however, for obvious reasons, and neither the publisher nor the author can accept responsibility for any problems of any sort that the reader may suffer, directly or indirectly, in following the advice in this book.

Acknowledgements

I would like to acknowledge the contribution of my colleagues Neil Bruce, Suzanne Cholerton, Julie Fyles, Jeff Idle, Bill Matthews, Elena Notarianni, Robin Mackay, Simon Penhall, Tom Rae, Nigel Scrutton, Colin Self, May Sullivan, Anne Willis and Kathy Willis. They have provided the inspiration to write this book and given considerable input as to its content. I am indebted to you all.

1 Bringing your technology to market

As the births of living creatures at first are ill shapen,
so are all innovations, which are the births of time.

Francis Bacon
Essay: Of innovation 1625

Introduction

We are living in exciting times. The pace of technological development in the 1980s and 1990s has been and will be heralded as a veritable revolution, whether this be in biotechnology, materials science, microelectronics or other 'high' technology fields. With this has come the need to incorporate novel information into products and services and hence, new industries have been born. This book focuses upon the biotechnology industry which, due to its fragmented nature, conveys unique characteristics that should be considered in commercial exploitation. However, it is hoped that the general principles that are described will be applicable to other fields and be of use to anyone wishing to take their technology to market. This book will be of relevance to academics at all levels, to university administrators, to entrepreneurs and to those wishing to enter the technology transfer business.

The technology gap

Molecular and cell biology is revolutionising many aspects of everyday life, allowing the coercive engineering of novel drugs, detection of genetic defects and disease states, making agricultural and industrial processes both more cost effective and friendly to the environment as well as making improvements in the quality of foodstuffs, amongst others. The US was first to realise the potential of this technology and some of the companies that arose in the early days are now familiar names, Genentech, Amgen and Cetus for example. In 1993 there were over 1200 US biotechnology companies, producing an entirely new range of medicines, foods, etc. Europe has been slow to explore the potential of these new technologies, for reasons that may in part be reflected by our cultural and historical caution. The UK biotechnology industry is probably 10 years less mature than that of the US. This is manifest in the income generated, the number of biotechnology associated companies that have been formed and subsequently 'gone public' and the amount of capital that has been invested. In spite of this, the UK academic community (often funded by the taxpayer) is highly productive. How can this success be translated into products and services that will contribute to the gross national product and individual wealth? Obviously a prerequisite for successful commercial productivity is a research community that keeps generating new technologies. However, in other areas such as the transfer of technology from academia to industry, a positive attitude towards the protection of innovation via patenting and a favourable environment for the financing of ventures and mechanisms to facilitate the introduction of new products, i.e. helpful regulatory and approval procedures, are all necessary. It is in the protection of intellectual property and technology transfer that Europeans must pay particular attention. If we get these right, then the potential wealth and enthusiasm that will be generated will no doubt ensure the necessary positive funding and regulatory environment. Since these areas are are fundamental to bringing your research to market, they will be the focus of this book.

The benefits to society of the commercialisation of academic research have been considerable, and the protection of the intellectual property that arises therefrom allows monetary proceeds from the research to be returned to the originating institute, thus ensuring more research, more jobs, more progress. US universities have been particularly adept at realising the commercial opportunities (e.g. the original Cohen–Boyer patent for genetic engineering that was granted to Stanford University), whilst the equivalent in Europe has been variable in quality and bereft of quantity. There is currently an unfair burden on the taxpayer, who is expected to pay for the research yet inevitably sees a fraction of its true value as a commercial social benefit, since the results are either patented in other countries or highly trained staff seek employment overseas. Some may say that a portion of this research is of such an esoteric nature as to be irrelevant. It is by increasing the interactions between academia and industry that we can reverse this trend and demonstrate that the research has true value to society (Fig. 1.1). It will then be much easier to justify funding. We must also look closely at our attitudes towards science in the market place, since this also presents a considerable barrier. Thus the first hurdle is not legal, but psychological. For example, mechanisms for commercialisation of research are often in place in UK universities yet British academics rarely use them; reactions range from general reluctance to hostility. In addition, there have been several examples of inexperienced academic staff entering into agreements with industrial collaborators only to find that they are severely compromised at a later date and have ended up with a 'deal' that severely under-compensates. Historically, there has always been an attitude that relates to

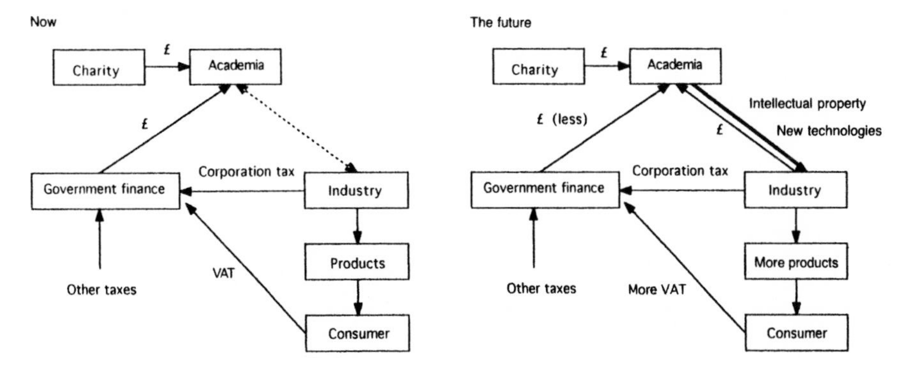

Fig. 1.1. Development of the UK research triangle: now and in the future?

the purity and superiority of non-applied research, which is partly due to both the erroneous perception of lower quality science in industry and partly due to the uncompromising selective processes in our education system. Such misperceptions must change if we are to assume our place in the technologically competitive twenty-first century.

There is a general perception that there is a shortage of finance for biotechnology ventures in Europe and especially in the UK. There is no doubt that compared to other sectors, biotechnology represents a high risk business and usually investors hope that a few extraordinary products will compensate for the large number of failures. The climate is now becoming less favourable, as too many new drugs are either not having the desired clinical effect or are causing unacceptable side effects. Derivation of a product from natural sources (e.g. gene cloning) does not mean that it should be easier to gain regulatory approval for it. Many companies have fallen in the rush to get such products into the market. In addition, several new 'biotechnology' drugs have been shown to have a much narrower range of applications than have been indicated in their respective business plans, with a concomitant reduction in market potential. Investors are cautious about the uncertainties in biotechnology patent protection, being aware of the immense drain upon resources that a patent battle would require, and indeed are concerned about the recent US legislation to limit the prices of pharmaceuticals. This may make European and Japanese companies more attractive in some respects, but the major demand for new therapeutic agents is undoubtedly of US origin. However, not all good ideas will necessarily involve therapeutics and thus require such stringent clinical trial and regulatory regimens; there are a great number of opportunities in related fields (e.g. research reagents, bioprocessing, hybridoma and other antibody technologies, bioelectronics, diagnostics).

Overall, the cash requirements in the biotechnology sector are likely to increase significantly in the next few years as the number of new companies increase and, for those involved in therapeutics, as clinical trials are performed and regulatory approval is sought. Delays in the latter processes result in a costly loss of patent protection time. There are four crucial factors that will contribute to the direction of any particular company, aside from the important availability of finance and regulatory approval, these are: (a) the size of the potential market and the probability of aquiring a significant share of this market; (b) the relative abilities of the management team; (c) the degree of technological advantage in the chosen field; and (d) the intellectual property position, i.e. whether the retained 'patent wall' is sufficiently strong to protect the company, especially in the US where the market is largest. Opinion is divided on the extent of competitive advantage afforded by protection of intellectual property, since there is discordance between the European, Japanese and US patent offices in both consideration of genetic modification and overall patent policy. The plethora of small companies that are all developing innovative new products but may lack sufficient resource to take these through clinical trials suggests that over the next few years there could be a significant number of strategic alliances, closely followed by merger and aquisition. For those individuals with access to novel biotechnologies and the will to commercialise, it is expected that there could be a concurrent significant financial gain. What is required therefore, is an understanding of how to transfer this new technology (patents, know-how, samples, products, information and processes) to the commercial environment, in a way that is acceptable to both the body which is paying for this to happen and that which owns the assets. It is not an activity that the average academic is likely to enjoy, since there will be a drain on time, a need for communication to people who are not directly involved in your field and the probability of only a small proportion of benefits from these efforts returning to the originator. However, it can be a rewarding if not cathartic experience, since the lines of communication that will be established will help focus your ideas and provide new avenues to funding.

The first objective of this book is to examine the parameters that affect the protection of your intellectual property and the mechanisms of the patent process. Second, consideration is given to negotiation techniques and the further development of ideas in the commercial sphere, including an indication of the difficulties that will be encountered in the valuation of the idea. Third, the licensing of the technology and the start up of a new company are examined, as are the financing and managerial issues associated with development of the venture.

2 So do you really have something of value?

Protection is not a principle, but an expedient.
Benjamin Disraeli.
In a speech made on 17 March 1845

The concept of intellectual property

Before proceeding to take your technology to the market place, you should first consider its protection. There is no point spending effort and time in entering a market when competitors could easily copy your idea and sweep you aside with a better integrated or superior marketing strategy. Legal protection may not stop this, but at least you will have some form of redress and a platform from which to build. Whilst the emphasis in this book will be placed upon the biotechnology industry, it is well to remember that the rapid fluxes inherent in this industry lead to a similar rapid evolution of the intellectual property position.

Intellectual property (IP) is the driving force of any business. Businesses acquire assets in the form of buildings, labour, etc. and these have a tangible value. However, what differentiates businesses is their ability to turn these assets into greater profits. This requires 'know how', the techniques, experience and awareness to turn the assets to best advantage. This is in part inherent in the expertise of the staff that have been employed but is also resident in the intellectual property. This is the protectable part of the 'know how' and is the source of added business value. The right to use intellectual property (an intellectual property right or IPR) is responsible for the establishment of markets, for generating loyalty, for controlling the industry and above all, for generating profit. IP is, of course, difficult to value and this in part reflects our human need to assess value in terms of things we can see. Success is often seen as an active production line, a series of oil rigs, a well equipped facility on a science park, i.e. we naturally equate profit to the successful deployment of the labour force and capital resources. However, the differentiation between products often lies in the concepts and designs, i.e. the trigger for one purchasing decision over another.

Intellectual property is the embodiment of commercial reputation, historical know how, goodwill and intellectual creativity. For the life scientist the latter is of extreme importance. There is a significant shift towards biotechnology-derived products and there are already in excess of 20 recombinant therapeutic products in use. The globalisation of biotechnology (enhanced by the encumbent power of the pharmaceutical industry) and the importance of advanced technologies to the gross national product of the country has given the biotechnology industry a high profile and a significant political lobby. Thus it is of prime importance that your IP be recognised and protected, both for your own benefit and that of the taxpayer.

For the life scientist, the legal protection of IP primarily involves protection of inventions. This is however, also important for designs, trade marks, artistic and literary items and computer programs. Intellectual property can be treated as any other form of property in that it can be legally mortgaged, licensed or assigned to another party.

Rights and duties

The ownership of IP means that there are various rights and responsibilities that simplistically can be considered to be equal and opposite. A right enables certain privileges, e.g. to use a particular vector or cell line and the corresponding duty is that everyone else should not infringe that right. Importantly, this right exists even if the infringer does not know of the existence of it. There are many compromise situations and exceptions; limitations may include compulsory licences, maximum times for exploitation and a consideration of competition law. With regard to the latter, possession of a patent could result in attempts to unfairly control a market. The monopoly position is open to abuse and both UK and EU law have provisions to control such behaviour.

What are the various forms of IP?

The pre-eminent form of IP protection for the life scientist is the patent (Fig. 2.1). The latter gives a 20 year monopoly right which can be granted for a new invention that is capable of commercial exploitation. The standards are rigorous and therefore the original patent application must be very accurate and precise, particularly with regard to the claims for which protection is sought. Patent rights are an ancient right originally granted by the Crown for certain priveliges, and indeed one of the first can still be seen in Salisbury Cathedral where there are 'letters patent' of Edward III granting one Robert, Bishop of Salisbury, free right of chase in Bishopesbere Wood in Windsor Forest, dated 15 April 1337.

Patents may be assigned or licensed to a person or company, and in order to use the information contained in the patent fees will usually be payable, typically in respect of the extent of likely use. This is embodied in a legal document,

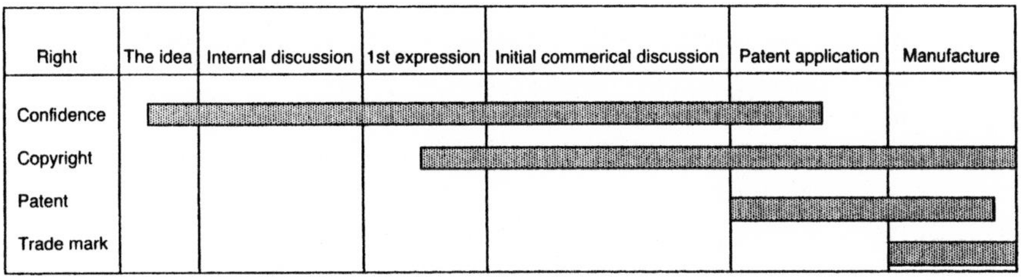

Right	The idea	Internal discussion	1st expression	Initial commerical discussion	Patent application	Manufacture
Confidence						
Copyright						
Patent						
Trade mark						

Fig. 2.1. Intellectual property rights.

commonly known as either an **assignment** or **licence**. Usually, patentable inventions are made whilst the inventor is an employee. Depending upon the employment contract, the inventor will always have the right to be identified as such, but the rights to the patent may belong to the employer. However, the inventor need not give up all hope and, if you should patent something that is of outstanding commercial benefit to your employer and there is no revenue sharing arrangement in place, you can apply for an award of compensation. Only you can decide if the internal politics of such a move are wise!

International patent harmonisation is proceeding rapidly, although there are still many inconsistencies across the world stage. In Europe we can expect a community patent to eventually become the dominant vehicle, but at present patents from individual member states are processed through their own offices and if required, the Munich based European Patent Office. For wider protection, there is an International Patent Co-operation Treaty (PCT).

Patents are supposed to be working documents. It is not sufficient to obtain a patent and then not use it, thus preventing others from exploiting the invention. For example, a new technology (e.g. a magic bullet antibody) may supercede an existing therapy (e.g. a drug treatment) that contains a significant element of vested interest (expected profits, etc.). It would not be acceptable for a company to buy the rights to the technology (i.e. be the assignee or a licensee of the patent) and then not use it, in order to protect existing sales. Such behaviour would be anticompetitive and prohibited. Thus, to prevent such abuse compulsory licences may be available three years after grant of the patent, with the terms decided by an independent arbitrator if necessary. In reality, this may be of limited use in biotechnology where the pace of technological advancement is high. Other forms of IP may be of less importance to the biotechnology sector but nevertheless, they will be outlined here to illustrate the range of protection that is available. If a product or idea is not sufficiently novel or inventive to warrant a patent it may be protected with a design right. As the business develops and your product attains market presence, you may consider protecting its image by registering a trade mark, a distinctive symbol or set of words that is identifiable with the product. A set of laws known as 'passing off' governs the misuse of registered and unregistered trademarks such that misinterpretation of the origin of the product can be reduced. If you have created a literary work, i.e. are an author of an article or a book, then you will always have the right to be identified as the author but the ownership may reside with yourself, your employer or another party. Copyright is not monopolistic and others will be free to create similar product.

For all forms of intellectual property, the law of confidence applies, i.e. it aims to prevent use of information for purposes other than those for which the right was granted. If the information is of an embryonic nature, i.e. pre-patent, then this could be a useful source of protection. However, misuse may be difficult to prove. If you were to give information in confidence on a method to grow a particular cell type, e.g. with a special growth factor cocktail, then it would be a breach of confidence to transfer this information within the receiving company for use on another cell type, unless permission is both sought and granted beforehand.

Having introduced the concept of IP and before discussing how to patent, we must consider the implications of government policy (*Realising Our Potential*, HMSO) upon the protection of IP within UK universities and research institutes.

Fundamental changes in the way we do science

Recent government policy is forcing a change of emphasis upon the UK academic community, towards an increasing involvement of our industrial counterparts. This is an attempt to increase national productivity and to make the UK more competitive with the US, Japan and Europe. It is often said that 'speculative' research is necessary in order to make fundamental break-throughs. However, the notion that more applied research (i.e. that capable of commercial exploitation) does not generate either generic or fundamental results is fallacious. With the appropriate managerial style, there is no reason why the research community should be less productive if there is to be a greater focus upon the applications of the research in question.

There are two major innovations that will become important during the next decade. Their success will determine whether this shift of emphasis to more applied research will, with hindsight, be seen to be a positive step. The first relates to international investment in the UK science base, a delicate balance between the commercial world and academia. This is sometimes referred to as corporate venturing. Whilst this is still in its infancy, there is a definite trend towards these associations. There are four ways in which such investment could be achieved: (a) a direct collaboration; (b) funding posts in a department, e.g. a professional chair; (c) establishing a unit on the university premises; or (d) by establishment of an adjacent research facility. The Department of Trade and Industry favours investment of this nature and we may therefore expect it to increase. For the universities and research institutes there are several advantages to such associations, beyond the obvious influx of funding and an increase in the number of posts, since it creates a centre of excellence in a particular university. However, at the moment it is very much a case of the 'strong get stronger' and the situation predisposes the style of research, i.e. inevitably there will be a shift towards research with greater applicability. For the company, the advantages are also considerable: (a) close links with the research community, i.e. a source of information; (b) access to human resources, i.e. staff that have been imbued with the corporate ethic; (c) improved options in the intellectual property arena; and (d) a soure of investment that could open a route to those markets which would otherwise be difficult to enter. For example, Japanese or other foreign companies may find this a useful first stepping stone into the UK/European market.

The second relates to the risk/reward argument and the effective development of commercially attractive ideals. There is a big difference between research and development. It is unfortunate, therefore, that they are often linked since there are a different set of skills and resources required for each. Many pharmaceutical companies consider that a competitive edge can be obtained by separating the two departments and concentrating the development resource, i.e. avoiding the inevitable dilution by research goals. Given the speed at which biotechnology is progressing, many pharmaceutical companies have realised that it is almost impossible to keep at the forefront of science at all times and in all areas. This is a major reason for strategic alliances between biotechnology companies and the pharmaceutical industry. The former provide the research leads for the development products of the future, which allied with the large financial resources necessary for both this and the marketing process, makes the alliance conceptually sound. In fact, if one was to extend this situation to its logical conclusion, one could envisage that the major pharmaceutical companies will eventually become 'development' houses that rely upon the research output from smaller specialist research companies. As these companies come and go in any particular subject area, the skilled staff will move from project to project in the search for new research avenues that will challenge their creative spirit. The close relationships that develop between the biotechnology and pharmaceutical sector are already showing in the US markets and indeed, are providing a useful filter for research quality. Unproductive or non-useful research products are not likely to find a customer and will thus be selected out.

The 'development' laboratory will require a very different culture to that of the research laboratory, with both different staff and attitudes. In this laboratory you can set clear targets and milestones. Rather than creative thought, systematic team and project based activities are required, together with cost effective project management. Development is on average nine times more expensive than research since it often includes the clinical studies. In the tough world of competition to put a drug on the market, pharmaceutical companies need to have a significant lead; the chances of a competitor making a mistake cannot be relied upon. So what does all this have to do with commercialising your research? Well, evangelical or prophetic it may be, it is inevitable that there will be a series of development laboratories set up to reduce the risk and take research projects closer to commercial reality. This will either be on an individual university basis (unlikely except for the largest universities) or perhaps regionalised to support development projects of several universities within a catchment area. When these laboratories are commissioned they will be an integral and efficient part of the mechanism for bringing your research to the market. It will not necessarily involve loss of ownership, but could go a long way towards reducing the risk of your venture, both for you and prospective clients. If non-commerciality is revealed (e.g. your bench top model cannot be scaled up to a useful prototype) then: (a) you will have saved a lot of your own time in the long run; and (b) you will not have compromised a potentially useful industrial contact. If the commercial validity is proven by a development project, then not only does the value (to you) of your product increase, but it becomes considerably easier to sell.

The hope is, of course, that UK universities will eventually augment their existing core competencies with an IP facility, becoming regional technology brokers as they: (a) hold on to a body of IP and judiciously licence; (b) provide facilities for start-up companies where appropriate; and (c) become a clearing house for consultancy services across a wide range of disciplines.

As these commercial relationships develop, so we might expect UK universities to generate revenue through larger, properly managed, contract research agreements and joint ventures, much as exists currently in continental Europe.

Changes in university attitudes to IP

We are undergoing a period of profound and fundamental change in the management of intellectual property in the UK. The days when the British academic community was a soft target for industry will soon be gone and the rethink will require considerable compromise on both sides. The adage of 'he who pays the piper picks the tune' will no longer apply as before, i.e. if industry pays for a piece of work, then it owns all of the resultant IP. The change will involve universities and research institutes holding onto their IP and building a portfolio of IP to justify their existence as centres of excellence. There are a number of conceptual changes, one of which is embodied in the nature of the commercial relationships between industry and academia. From the industrial side, the partner is likely to be somewhat irritated by the 'research grant' mentality of many academics, i.e. the tendency to cost projects on the basis of a research grant rather than on the basis of a commercial proposition. To the defence of the academic, they have probably never known anything else and more to the point, how would they find out about it? Who provides relevant training? The problem is compounded by the often belated and beleaguered attempts of university administrators and technology transfer offices to redress the situation, resulting in an incoherent front being presented to the sponsor. No wonder there is often an air of disdain from commercial partners regarding university research. Fortunately, the government has instigated a major period of change whereby not only is the policy on intellectual property going to change, but the whole basis of university research is going to be put on a more applied and commercial footing.

Section 39 of the Patents Act 1977 states that the ownership of the IP rests with the employing organisation. If inventions are made by the academic staff during the normal course of their duties then the university will have a justifiable claim to ownership. If funded by a UK research council, then the rights to IP probably rest within the institute in which it is generated and indeed, many UK universities have a mandate to exploit the results of research council funded research. In the development of a research group, a considerable amount of public finance will have been used for purchasing capital equipment and building up a base of intellectual property, sometimes with the participation of many personnel over several years. It is this unwritten base of know-how that will provide the launch pad for any new intellectual property and perhaps a patent application. Any group of this nature will be producing intellectual property in several areas simultaneously and over time.

Therefore, there needs to be a concerted and integrated approach to the management of such an intellectual property portfolio. This is particularly important if you are going to be dealing with several sponsors, some of which may overlap in product areas. It would indeed be unwise to allow all the intellectual property that arises from a project to reside with a single sponsor, since it could severely limit the research potential of the laboratory, the interactions with other sponsors and perhaps even the long-term integrated research programme of the institute. Ideally, the university should attempt to hold onto all of the intellectual property and grant specific licences to use aspects of the technology in defined technical areas and geographical regions. If there will be considerable support for a research project to be carried out in the university, then there will always be an argument that the sponsor should have total and exclusive rights to the technology. This should be resisted and the university should negotiate to retain all intellectual property rights, with the intention of building up a marketable portfolio of the same. The sponsor should be offered first option to use the information within a specified period of time. In this way, if the sponsoring company does not wish to use the information, e.g. because its research directions have changed in the years since the initiation of the programme, then the university will be free to licence to another parties or carry on using the information without prejudice.

If the inventors are employees of the university then there may be a revenue sharing arrangement, subject to the provisions of the Patents Act 1977. There are, however, usually predetermined guide lines, for which each individual university and research institute should be consulted. These are designed to provide an incentive to inventors to come forward with their new technologies. There may occasionally be a contribution from non-university employees such as Ph.D.

students; in such cases you should ensure that the position is clarified and they either sign the main IP agreement with the sponsor or assign the rights of their invention to the university.

Patenting your technology: what is required for a patent to succeed?

During the last decade, the biotechnology industry has witnessed a large increase in the number of patents applied for, which has created both considerable debate and the evolution of new precedents in several areas. The objective of the next section is to introduce aspects of both patent law and the patenting procedure. Subsequently, the patent application procedure will be discussed as will topical precedents such as the circumstances surrounding protection of human DNA sequences and novel life forms.

☛ Preliminary considerations

There is always a search for new products that allow a company to gain competitive advantage. However, no one within the organisation can gain from a failed product and in general companies are risk averse, thus making it difficult for new products to gain acceptance into the product portfolio. Most 'new' products evolve from those already in existance and are thus innovations rather than inventions. By definition an invention is unfamiliar and full of market risk, since it is not known whether sales can actually be generated. It is, therefore, absolutely essential to get your idea to the right people and in the right way, i.e. the way that is most consistent with what they are familiar with. There are a number of areas where the inventor and/or technology transfer office may make early mistakes, and you should enter into a full and frank examination with a relevant third party to assess the following points. The invention must be original and therefore you should carry out a thorough search for similar products and for relevant scientific literature or patents in the field. If you consider the product to be original, then you must ensure that it is saleable, which necessitates thorough market research. Inventions usually solve a problem, so make sure that the latter has been thoroughly analysed and that you have an optimum solution, preferably with a product that is simpler than pre-existing technology and fulfills an unmet need in the market.

There are a particular set of problems that your technology transfer office is likely to face. In the initial appraisal of the technology it may be necessary to limit the expenditure in the project until the idea has crystallised. Should this budget be spent upon a patent search, actual patenting, a market survey or production of a prototype/sample? Furthermore, is the inventor realistic in commercial expectation? Nearly all inventors will be totally convinced that their invention is a world beating opportunity and as such they may require a little education in commercial realities. Some inventors use patents for political gain and competition between academics to patent is likely to increase, especially as one expects industrial liason to eventually contribute to the Higher Education Funding Council (HEFC) rating for UK universities. In addition, it is important to target the right spectrum of potential customers who may not necessarily be the market leaders.

☛ The search for prior information

The first recourse is a patent search, whereby you will identify prior art that is already in the public domain; this is the most efficient method for determining whether your idea is original. Patent libraries have access to all patents that have been published and you can obtain the relevant information by either searching yourself, asking a patent agent to search or using the Search and Advisory Service provided by the Patent Office. The latter will cost up to £500 depending upon the search, a modest outlay that could save you money in the long run. If you choose to do the search yourself, it would be sensible to consider the use of the patent librarian who may charge a modest fee to cover on-line computer time. Such persons will also be able to help with identification of the patent classification that best fits the invention. A number of on-line searching services are available for subscribers and your institute or company may be able to assist. For biotechnology inventions it will probably not be necessary to search more than 15 years into the patent literature, given the recent emergence of the subject. The choice of words for a patent search should be as you would normally use for a scientific literature search, but be aware that you may miss some patents that are relevant. The search performed should also include foreign patent applications. Patent searches do not prove originality and the inability to find prior art does not necessarily mean that you should patent immediately, since there will also be commercial considerations.

The second recourse for information is the scientific literature. We all have access to on-line literature search facilities and it is relatively easy to scour the literature using a series of key words. The reference section in the library will also be of assistance where general texts such as Dunn and Bradstreet or Kompass can indicate manufacturers and suppliers of like products. More specialist libraries will have biotechnology directories that can provide useful information on suppliers. There are also several databases, which include press reports (newspaper and journal cuttings), that can often give information on industrial activity. Similarly, the catalogues of companies in the sector should be obtained and scanned.

☛ Assessing market demand, utility and commercial potential

The next important criterion is whether the idea has commercial value, which must be determined before patent costs are incurred. This exercise should be carried out by the technology transfer office, by the marketing staff in your 'start-up' business or by yourself. Only recourse to the latter if you are sure you can be objective. In order to collect the market information, attempt some preliminary market research, broadly trying to answer the questions indicated (Fig. 2.2). Much of the information required will be difficut to find, quantify and justify. However, in order to pursuade a prospective client to purchase the technology or convince yourself of an adequate financial return, then the issues indicated must be addressed and not ignored, simply because they represent difficult questions.

For market research to be most effective the first approaches must be made to potential manufacturers, as they will have to invest considerable financial resources in taking the product to market and will therefore have a lot to loose if sales are not forthcoming. Their opinion will be more important than those of the consumers, since the latter cannot be forced to purchase and do not have a vested interest no matter how much they say they like the product. A thorough market research campaign can take several months to complete and should take resource precedence over production of a prototype, since the results of the former could well alter the nature of the final product. For example, the discussions that you may have with the manufacturers may cause you to either modify the design of the invention or adjust the market niche that you consider appropriate. Market research should be considered an internal matter, as professional market research companies are both expensive and may not have either the expertise or willingness to take on a biotechnology associated project. To enter the export market, a considerable amount of information will be required and this can be obtained by enlisting the services of the Department of Trade and Industry (DTI) export initiative, which will be discussed later.

Many products fill a gap in the market that may otherwise remain hidden, i.e. the potential customers often do not yet know they 'need' the product and remain to be convinced. Alternatively, there may be a latent demand such that when your product becomes available, its advantages over existing products are immediately obvious. How does one identify the demand for a product? Your technology

Questions for market research	Comment
What is the market for the product?	Delineate as many areas as possible and prioritise them
How extensive and valuable is each market?	Categorise them as growing, static or declining, quantitatively if possible
Is there a definite niche to exploit?	Is it novel, more efficient, cheaper, etc.?
What are the competitive products?	Define, source information on prices, etc.
What is a realistic market share?	Quantitate and allow for both uptake and competitive responses
How many units do you expect to sell?	Also try to estimate the frequency of repeat purchases/replacements
What is the expected selling price?	Determine how much the market would pay for the product

Fig. 2.2. Questions for market research.

transfer office may have a considerable body of annual reports and company literature, together with expertise in the required subject area. Failing this, telephone these companies directly or obtain information on microfiche from Companies House. Additionally, you may find that a well-constructed questionnaire has value, but beware of bias in the replies and your analysis of them, since of your original mailshot only a few will reply. Considerable skill and experience is required in order to design questions which will give you answers that can be usefully applied. In order to facilitate responses, try to include a stamped addressed envelope. Another route is direct personal contact or the construction of a network; without compromising confidentiality it is surprising how a jigsaw can be built from publicly available information. Obtaining information from companies is extremely difficult whichever route you choose, and indeed your enquiry may lead to you receiving deliberately misleading information or the triggering of commercial ideas in that company. For example, the entirety of your idea may not be marketable but part of it may be. Inevitably market research will become a matter of informed guesswork and over a period of time you will intuitively build up a picture of your proposed market. In doing so try to ensure that there are at least some verifiable facts upon which to base your assumptions, sufficient to ensure credibility during later stage discussions.

The press has a surprisingly large amount of company information and in particular the *Financial Times* and *The Economist* often have special features on specific industries and/or national markets. The City of London business library is a remarkably complete and useful source of company and market information, often containing copies of limited edition market reports. In addition there are several databases that contain press cuttings, together with original references. In all cases you should attempt to obtain information from the primary source and not the edited versions that are usually found in the newspapers. Most articles quote the original sources and these should be found.

A significant amount of information can be obtained at low cost from company literature or from Companies House. For example, you could ring the company in search of a price list/brochure etc. or obtain the details of shareholders/accounts from the latter. Providing you feel confident at analysing the accounts you will be able to ascertain the financial stability of the company. Do not spending valuable time on those that may become insolvent or have little cash for investment on new products.

The next step is to produce a prototype of the invention if this is indeed possible. This could be a working model, a batch of enzyme that can be subjected to extensive quality control procedures, data on a new diagnostic kit showing improvements over the competition, etc. In essence, your prototype should be sufficient to show that the invention will actually work in practice. This could be quite simple at first as you concentrate upon functionality, but it will eventually need to be akin to the proposed final product when you are trying to sell your idea to commercial concerns. There are several methods by which you can obtain help with the manufacture of the protoype, perhaps via the technology transfer office that could arrange for you to perform the necessary experiments in an existing university or institute, or you could contract out to a regional innovation centre upon advice from the DTI. These data so obtained will help answer the next most important question, i.e. can the product be turned into commercial reality at a realistic cost? This inevitably means calculating a unit manufacturing cost that can be quite difficult to the uninitiated but, on the other hand, if it can be based upon realistic figures it will enhance your credibility and hence also your chances of a sale to prospective clients.

Launching a new product will involve a considerable amount of money and development grants that are available to companies are not available to individuals. Hence it is usually more cost effective to work with a corporate partner such that you or your institution share the risk. In doing so your rewards could ultimately be greater, since setting up a manufacturing capability is both extremely difficult and will take a great deal of time and effort, neither of which you may wish to afford.

So now having assessed your patent for validity and commercial potential, what are the criteria which surround the award of a patent?

Requirements for the grant of a patent

The law of intellectual property has special provisions for inventions which conform to a given set of high standards and these result in what is commonly known as the patent. It is the main form of protection used in biotechnology today and as such, considerable discussion will be devoted to the intricacies of the patenting process. A patent is an agreement or bargain between the state and the inventor or owner. In essence, the exchange is a monopoly for a limited period in return for public disclosure of a technical description that is sufficient to repeat and use the invention.

One particularly important point is that the grant of a patent does not confer the rights to use a patent, but the ability to prevent others from doing so for the remaining length of time for which it is in force. This should give a considerable boost to your research and development effort, since your interests will be protected. The Patents Act 1977 provides for both assignment and ownership of patents by third parties. A patent is not like a scientific paper in that the consequences for not fully disclosing your technical details can be severe, i.e. loss of the monopoly. The patent system would not work if every technical variant was protectable and hence the concept of prior art must be defined. Prior art is the total body of information that is publicly available before the filing date of the patent and could be published papers, meeting reports, computer databases and so on. The grant of a patent is governed by the provisions of the Patents Act 1977, which requires that the invention: (a) is new (novel); (b) contains an inventive step;

and (c) is capable of exploitation in an industrial or commercial context. Careful consideration of these criteria could save you both time and money and they will now be discussed in detail.

☛ Is your invention really new?

In patent terms this means that your idea or invention must not be available to the general public or scientific community. The Patents Act 1977, Section 2 (1) has a rather cryptic phrase to describe 'new', being that the invention must 'not form part of the state of the art' in order to qualify. **State of the art** is further defined as any product, process or information that has been made available to the public before the priority date of the patent application, including written or oral presentations (Section 2 (2)). This may seem contrary to the scientific ethos of many of us, as few can say that they are not usually delighted to pass on information to colleagues and in many instances, to the press. If you think you have something worth patenting it is wise to pause and reflect for a moment on the consequences of your disclosure. If you are in the biotechnology industry there will of necessity be times when you need to reveal your inventions and general base of intellectual property to say, new investors or potential customers. Similarly, the academic wishing to capitalise upon an invention will need to attract interest from potential commercial sponsors. Make sure that you hold these meetings in private and obtain a properly constructed and signed confidentiality agreement (often called a non-disclosure agreement or NDA) from the participants, preferably before your first meeting. Any patent agent or solicitor will be able to draft you an agreement for a small fee. Information obtained in breach of such a confidence is considered to be illegally obtained and will not be considered when your case for novelty is discussed, providing that the priority date of the patent is within the six months following the illegal disclosure. Discussion of results within your research group, e.g. at laboratory meetings, are not generally considered to be public disclosures since the meetings are closed and each member should be under an obligation of confidentialty. It would not, however, do any harm to state this to incoming members of the laboratory, if you are researching in areas that may generate novel technologies.

A further constraint upon the novelty of your invention is that it must not have been anticipated by any person, which includes someone 'skilled in the art', i.e. a reasonably competent colleague whether this is by general use or a previous publication or patent. The notional person who is 'skilled in the art' requires some explanation; the person is hypothetical and not selective in his/her knowledge. This person knows everything there is to know about the field, providing it is in the public domain. This implies that this person must have read all of the available literature, both scientific and patent. Furthermore, this person must be able to make routine developments, but must not express either ingenuity or imagination. To many scientists this seems heretical, since their training is one of discovery and invention. It should also be noted that the hypothetical person who is 'skilled in the art' could also be a team of researchers.

Another potential danger is self-anticipation by either publication in a journal or giving an oral presentation before a patent has been filed. This includes *any* meeting abstract. Attempting to speed up the process of publication by submitting the paper before filing the patent is also dangerous, since you do not know who the referees are and, furthermore, they are generally not required to sign a confidentiality agreement with either yourself or the journal in question.

☛ Do you have an inventive step?

The next test to be applied is whether the invention is obvious to one 'skilled in the art'. The 'skilled worker' test must be applied to someone who is non-inventive, otherwise all inventions would be obvious. Inventiveness has to be decided in retrospect and an important factor may be the degree of commercial success of the invention. This is because if the product(s) fills a market niche, then this becomes an indication of non-obviousness, i.e. one must ask why someone did not do it before? However, the reverse may not apply, since a product(s) may be inventive, but not necessarily a commercial success. Considerations for inventiveness will also include an appraisal of the size and nature of the problem that the invention seeks to solve, and whether there have been (and how many) unsuccessful attempts to solve this problem. The solution to the problem will also be assessed in order to determine if the invention is contrary to conventional teaching and practice.

The notion of the inventive step is indeed difficult to define since it relies upon an objective decision from a skilled person who knows the prior art. It cannot be an obvious development from the existing prior art, otherwise routine technical developments would find their path halted by the patent system. A discussion with a patent agent will help resolve the subjective versus objective argument.

☛ Does your technology have industrial utility?

The requirement here is for a product that is commercially and practically useful in an industrial setting. In essence, this distinguishes the patent from other forms of intellectual property such as copyright. Patent law in general excludes computer programs, mathematical and scientific models, business plans and methods for the treatment of either humans or animals by surgical procedures, diagnosis or therapy. However, material compositions or substances used in such treatments are usually not excluded. As an illustration, a topical cream to cure a *herpes simplex* infection would be patentable but the method of applying it would not necessarily be so. The aim of course, is to prevent hindrance to the clinician.

Handling the patent

Before we consider the application procedure itself and the costs involved, we will first turn our attention to areas of patenting where there is often confusion. These will include the issues surrounding who owns the patent, inventorship, record keeping, etc. For more detailed analyses and examples you are referred to one of the many excellent books on intellectual property.

☛ Inventorship and ownership

The inventor on a patent must be distinguished from the owner of the invention. A patent may have joint inventors and always have the right to be identified as such. Indeed, failure to do so could halt the application process. Inventors may in some cases also be the owners of the patent.

The owner is that person(s) or business entity which has the right to work the patent. An invention made under a contract of employment will belong to the employer. This means that if the invention was made in the course of the employees normal duties, those specifically assigned by a person in a senior position or if there was a special undertaking to further the company's business, then that invention belongs to the employer. If there are several owners, each co-owner can perform acts in respect of the patent without the consent of the others, excepting licensing, assignment or mortgage. If one owner dies, then that share passes into the persons estate and not to the other owners.

Who should be cited as an inventor on your patent? There have no doubt been many discussions on this point and doubtless there will be many more in the future. It is hoped that the following discussion may help clarify some of the issues. First, under UK law, the inventors are 'the actual devisors of the invention'. This refers to those who had the original idea, specified and, in some cases, executed the subsequent project. This does not include skilled technical assistance, except where an associate followed a path of their own, maybe against instruction or conventional teaching, and subsequently produced data in support of the patent application. In the latter case, it would be difficult to exclude such data. Second, there is little legal room for choice in the matter of inventorship. Contributions must be justifiable and this does not necessarily include superiors in the organisation (i.e. your boss, laboratory head or head of department/institute) who have provided advice, encouragement, finance or a facilitating role. Similar arguments apply to the co-authors on a scientific paper. You may wish to recognise contributions of technical assistants, provision of reagents and departmental seniority in your published papers, but this does not constitute inventorship for the purposes of your patent.

What do you do if your head of laboratory (who may not be an inventor) demands inventorship on the basis that the work was performed in their laboratory or institute? Legally speaking, this is not a valid reason for claim of inventorship, especially if the intellectual input has been minimal or non-existent. The politics of these situations can be very complex and there are often issues regarding references for future employment. Indeed, some attempt to reduce the number of moving parts by offering junior 'inventors' a nominal sum, e.g. £1000, in return for assignment. In the final analysis, you may find it is up to you to balance the pros and cons for your own situation.

☛ The importance of a proper notebook

A clearly written notebook is perhaps the most valuable result of your laboratory efforts. In the US, any dispute over who was the first to invent a particular process will have the respective lawyers arguing over the notebooks on a page to page basis; any lack of completeness or clarity could cause the case to be lost. UK filings operate on a first to file system, but recent changes that are a result of the General Agreement on Tariffs and Trade (GATT) Implementation Act in the US, means that from 8 December 1995 (1 year from enactment) it will be possible to prove a date of invention in the US which relies upon experimental work performed in any country that is signatory to the 8 December 1994 World Trade Organisation (WTO) agreement with the US. This a great step forward in the harmonisation of intellectual property law, but will only be of value if non-US researchers adopt the US requirements for proving the date of invention.

There are some simple tips to follow that may seem laborious and a bit of a nuisance, but which could either pay dividends or prevent a highly uncomfortable situation in the future.

1. Record all details of the experiment, not in excessive detail but in such a manner as to make your experiments easy to follow and repeat. This should be achievable by someone with a reasonable degree of skill in the field of interest, e.g. molecular biology or cell biology. This should include sources of reagents, details of growth media, buffers, methods, etc.

   As and when you draw a conclusion, it should be written down in a clear and concise manner. It is also important to record any ideas, hypothesis or projected plasmid constructs for example. Avoid abbreviations or code names or coded samples that you alone will understand.
2. Date and sign every page of the note book as and when you finish it. It should be witnessed as your work by a second person, who is not a co-inventor, but is in a position to understand your work. Full details on inventors and witnesses should be kept (e.g. updated addresses) since these may be required for rewards to be distributed or for court proceedings.
3. Computerised data storage is again a dangerous method for patent purposes and, if possible, you should print out your data once a week, annotate, date and sign it as before, with a co-signatory.
4. Any permanent records such as computer printouts, gel photographs, cell

photographs, protein purification profiles, western blots, etc. should be bound into a permanent note book. Avoid the use of ring binders and try to keep one book per project.

5 The notebook is the property of the laboratory. Ideally there should be copies made and stored in a separate place in order to avoid any affects of fire damage, etc. Some larger companies even store notebooks on microfiche. If you are a laboratory head, it is important to ensure notebook discipline regarding writing up, signing and leaving the book in the laboratory after the employee has left. This includes postdoctorals, no matter how independent. There is, of course, an opportunity for people to keep their own records after leaving, but this must always be made subject to confidentiality arrangements.
6 It is also advisable to send a copy of your notebook to a US counterpart or sister company, since this may be relevant to interference proceedings.
7 Notebooks should be numbered and handed out to researchers by the laboratory head; information should be recorded in indelible ink, without blank pages or pages removed. Any crossing out should be dated and initialled.

☛ What other records should be kept?

Accurate records of each project are highly important and you should keep a separate file for each project that involves IP, to contain: all letters sent and received, arranged in chronological order; any relevant scientific papers, leaflets, patent applications, specifications from catalogues or otherwise, price lists; a written record of phone calls and meetings; photocopies and originals of any facsimilies received and sent, plus evidence of sending; photocopies of reference material, newspaper cuttings, etc. Furthermore, only send copies of originals if requested and keep an abbreviated file of relevant letters for quick reference at meetings. Also, make a quarterly note explaining what you have achieved and what you expect to achieve, with approximate timings, and if a third party is involved, your expectations of them.

☛ What constitutes a prior disclosure?

This is an issue about which there is much confusion and some clarification will be given. Remember, public disclosure prior to the filing date may prevent grant of the patent in many instances. The following instances apply. (1) In patent law, a public talk or poster presentation at a meeting is sufficient to constitute prior disclosure. A public talk includes the now prevalent small closed meetings of 20 participants. (2) You should not submit manuscripts for publication before you have filed a patent. First, there is no legal obligation of confidentiality on either the journal or reviewer. Some reviewers do not take their obligations seriously and may ask a postdoctoral fellow to review the paper because they are too busy, or they may pass the submitted manuscript around. Second, your publication may be accepted for a journal and your rush to file may leave a patent application with errors or omissions. There is a lot of concern in many academic minds regarding the perception that patenting could delay or even suppress publication. Similarly, it is not advisable to submit an amino acid or DNA sequence to a public database before filing as this is also considered to be publishing.

The other even more difficult area is that of the grant application. Again, it is not advisable to disclose inventions in such documents, but this may be almost impossible since you will need to pull out all the stops in order to get the grant awarded. In this case, you should simultaneously work on both a new idea for a grant and the patent application. Make sure you file as soon as you are satisfied that you have the three necessary criteria in hand. Premature patenting could lead to excessive costs and the precise timing should be discussed with your patent agent. In many non-US countries, the patent will be awarded to the first individual(s) to file, so it is important to get the timing right. In the US the rights usually lie with the first to invent. If you are contesting a US patent application, then you can see that the correctly written up notebook is essential.

☛ Insurance for IP

Intellectual property insurance pays for the legal costs associated with pursuing infringers, by whatever means you or your patent agent or solicitor consider appropriate. Should you unintentionally infringe third party rights it will also cover your defence. A number of fees can be covered, such as those that relate to patent agents, expert witnesses, costs of appeal, etc. A commercial insurance broker will be able to advise you on suitable policies, which are, in general, not expensive. It is essential to insure the IP before publication of the material, since infringement could occur immediately upon publication and any damages you could receive will relate back to the date of publication.

☛ Your rights as an 'employee' inventor

The Patents Act 1977 provides for inventions made whilst you are employed such that the invention will belong to the employer if 'it was made in the course of the normal duties of the employee or in the course of duties falling outside the normal duties, but specifically assigned and the circumstances in either case were such that an invention might reasonably be expected to result from the carrying out of these duties' (Section 39 (1) (a)) or 'the invention was made in the course of the duties of the employee and at the time of making the invention, because of the nature of the duties and the particular responsiblities arising from the nature of the duties he/she had a special obigation to further the interests of the employers undertaking' (Section 39 (1) (b)). Clearly then, for most universities, research institutes and companies the patentable material will belong to the employer. However, the Act does provide for provision to the employee of a fair share of the benefit which the employer has derived from a granted patent, taking account of Section 41 (4): (a) the nature of the employees duties and remuneration; (b) the effort and skill devoted to the invention by the employee; (c) the effort and skill devoted to the invention by others (advice and assistance of non-inventors); and (d) the contribution of the employer (advice, facilities, provision of opportunities by virtue of management and commercial activity).

☛ What you can and cannot do with your patent

Patents may be assigned, mortgaged or licensed to a third party, providing these transactions are documented in writing and signed by both parties to the transaction. Licences to use the information in the patent or in the patent application can be granted, with the possibility of sublicences being available thereafter. The concept of the exclusive licence is discussed in Chapter 6, but at this point we should consider that such a licence excludes everyone, including the applicant or owner, from any rights to use the information in the patent to which this relates. If you give an exclusive licence, then the **licensee** will have as much right as you do to chase potential infringers.

All exclusive licences, assignments and mortgages must be registered at the Patent Office to: (a) protect the new proprietor's or licensee's rights to obtain redress for infringement; and (b) to give proof of title. If a **licensor** suppresses an invention for commercial or other reasons, then a compulsory licence may be purchased. For example, this would apply if you licensed your patent for a new drug delivery system to someone already in the field (quite likely) and then they did not work the patent to maximum advantage or did not meet demand, perhaps because they already had a significant series of products in this area. In such an instance, three years beyond the grant of the patent, compulsory licences may be allowed by the comptroller of patents.

☛ Infringement of your patent

The rights granted with a patent are strong and it is a necessary consequence that the action that can be taken for infringements are equally strong. The patent specification must be clearly and precisely presented, so that the court can decide whether variations to your invention are an infringement of your rights. If your specification is too narrow, then variants that could considerably affect your market could be made in a lawful manner. If the specification is too wide, then the application could be rejected.

It is an infringement for someone to supply another with the means to exploit a patent if the parties involved are not the proprietor or a legitimate licensee. If the patent is for a new product (e.g. a new cytokine) then it is an infringement to make, keep, dispose of or use it. Similarly, if the invention is for a process (e.g. how to make the cytokine) then it is an infringement to: (a) dispose of, import, keep or use a product produced using the patented process; and (b) use the process knowing that it is without the consent of the proprietor. Exceptions and precedents abound in patent law, since the circumstances and issues involved are rarely straightforward and there is a legal debate as to whether it is sensible to approach the patent specification in a literal fashion. Thus, with the aid of your patent agent: (a) define those elements that are essential to your invention; and (b) try to consider the variants that are closest to your product or process. In drafting the claims you should ensure that the scope of the application is not set too narrow, thus reducing the scope for potential infringement.

If you allege infringement or indeed are the defendant, then there are several defensive positions that could be adopted. Again, these should be taken only with the advice of your patent agent. The first line of defence may be to challenge the patent itself, perhaps on the ground of lack of novelty. It is also possible to challenge the proprietors' rights in the patent, expiration dates, etc. Similarly, a defence may be that the alleged infringement of usage was for experimental or private purposes only, i.e. not for commercial reasons. A number of other circumstances have been used.

Companies with significant financial resources (and teams of legal advisors) may threaten small companies and/or single inventors/universities with infringement, knowing full well that they may give up to avoid costly legal proceedings. If you consider that the threats are unjustifiable, then a declaration of such is possible, which could result in both damages for financial loss as a consequence of the threats and an injunction preventing the threats from continuing.

Remedies for patent infringement are fairly obvious, to include some or all of: an injunction, declaration of infringement, giving up products and/or documents relating to the patent. Damages may be awarded and calculated on lost sales or actual profits derived from the number of products sold.

☛ Assignment

As with other personal property, IP rights may be assigned to other persons for payment or as a gift at which point the recipient (assignee) acquires all the rights that would otherwise be attributable to the previous owner. In assigning IP rights, both parties (the assignor and assignee) should sign a written agreement that effects the transfer. For patent assignments this document must be lodged at the Patent Office. If the price at which the IP is sold exceeds £60000 or if the licence is irrevocable and exclusive, then a stamp duty of 1% of the purchase price will be payable. Whilst there are no immediate penalties for non-payment, a liability to penalties and accrued interest may arise if the formalities are not competed within 30 days of assignment. In addition, unless the duty has been paid, the document cannot be used as evidence of assignment.

In some cases if you are commissioned to do a piece of work, the assignment may be made before the right exists. Such arrangements are often made by publishers or software houses.

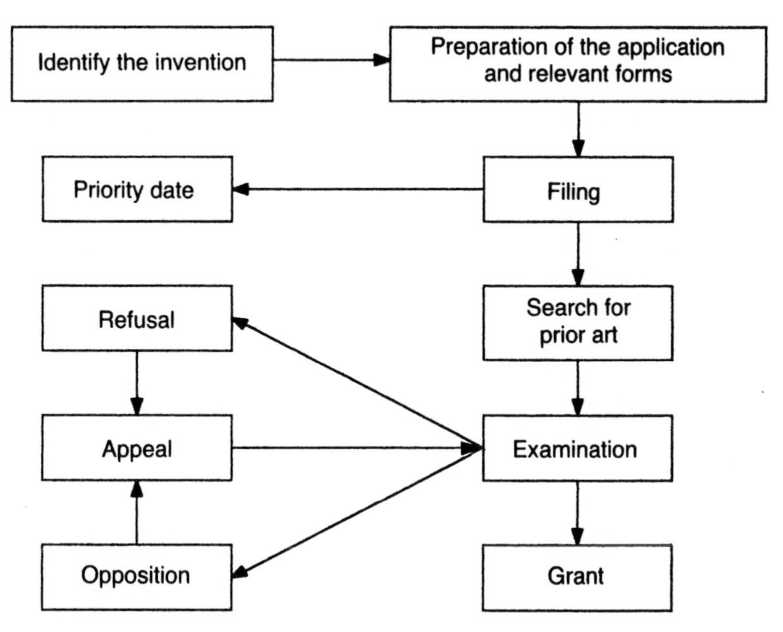

Fig. 2.3. Key steps in a patent application.

Preparation of the patent application

If a patent is to be commercially useful, it needs to be carefully drafted and a patent agent is essential. The purpose of this section is not to provide a 'do it yourself' guide, but to highlight the procedures and processes that you are likely to encounter (Fig. 2.3). Most of these points are applicable to UK patents, but laws between countries vary considerably and this should be considered in the light of professional advice. There are three parts to the application: (a) the description; (b) the claims; and (c) the abstract.

The **patent application** or **specification** contains a description of the invention, the nature of the claims and necessary illustrations. The description begins with a short descriptive title, which ideally should not reveal the nature of the invention or contain any abbreviations. Immediately after the title should be the background to the invention and the reasons why it was made. The description should then be written in clear concise language (English for a UK application) and contain sufficient information for someone 'skilled in the art' to carry it out, solely from your description. It should include figures, tables and explanatory drawings where these augment your case. It is absolutely crucial that you disclose as much relevant information as you can, since it cannot be added at a later date and may well determine the validity and commercial utility of your patent. If there are several forms of the invention, then these should be described at this point.

The **claims** must: (a) define distinct technical features; (b) relate to the description; and (c) not describe merits, rewards or advantages over other inventions. They must embody that which is about to become your property as distinct from someone elses property. Hence the claims must be written so as to clearly delineate the territory that you believe to be yours. It is common practice for claims to be quite broad, narrowing as the application proceeds through the examination process. This is to ensure that your application has as broad a scope as possible. The claims cannot be widened post filing and it would be highly unfortunate if you missed a key point.

A patent application must also contain an approximately 150 word abstract, submitted on a separate sheet. As with a scientific paper, the abstract should be a concise summary of the invention, for which you may refer to a figure from the specification if you so wish, but you may not include actual figures in the abstract. The Patent Office may amend the text of the abstract.

There are a series of requirements for the presentation of the figures in your patent application and rather than recount them here, the best option is to contact the Patent Office for details.

☛ The first stage: making the patent application

The first step in making your patent application is to obtain the form, 'Request for Grant of a patent' (form 1/77) either directly from the Patent Office or from your patent agent. If you are not using a patent agent, send the completed form, filing fee and the application (description, claims and abstract) to the Patent Office. For extra security, you can take it there personally or use registered postage. Alternatively, the patent agent will deal with these issues on your behalf. Once the office has received these documents your patent is said to be filed and you have that all important filing date. The significance of this date is that it gives you priority over those who may file later. The application will be published within 18 months of the filing date in most countries, notably excepting filing in the US. In this case, the information contained in the patent is only published upon its grant.

In the year post filing, additional patent applications may be filed that either modify the original or present new and related data. This new data can be consolidated with the earlier application (the first filing) and may assume the original filing date in a process known as claiming priority.

Upon filing, the next stage is to: (a) ensure that the claims and abstract are filed; and (b) submit the Patents Form 9/77 that requests a search (for which a fee is payable) and a preliminary examination. Ideally you should do this at the time of filing, although there is a time limit of one year from the filing date. The Patent Office will then send you the search report. Within the first year you must determine if your technology has commercial utility and the patent system is designed to give you this breathing space. Ideally you should use this year to find a licensee and enter an agreement that includes cover of the future patent costs. Any private expenditure on patent costs is not tax deductable and as such, it makes good commercial sense to arrange for the licencee to pay if at all possible. However, in the majority of cases the fees will be paid by the employer and there will probably be a budget to cover such matters. Once your 'commercial mind' is set you may then consider foreign patent applications, using the UK priority date.

In many instances you may wish to protect your potential innovation by gaining a priority date but still be unsure as to its commercial utility and, therefore, also be unsure as to whether it is sensible to make the necessary financial outlay. This period may also be one where you are seeking finance for the project. The minimal filing is as before, form 1/77 and the description. At this point, there is a year to decide upon how to proceed: (a) to **refile** with new information claiming the previous filing date; (b) continue with the application as is, by ensuring that the claims, the abstract, the search fee and form 9/77 are submitted; or (c) do nothing and allow the application to lapse.

If you are the applicant and not the inventor, then the Patents Form 7/77 will also be required, which will ask you to disclose the identity of the inventors and justify the position of the applicant. This form is required within 16 months of the priority date.

☛ The second stage: preliminary examination and search

Upon arrival at the Patent Office, the application will be checked to ensure it meets the formal requirements and if it does not, you will be given a chance (within a defined time period) to rectify the situation. In addition, a technically qualified person will undertake a search of all patent documents. You will then receive a preliminary examination and search report concerning your application, possibly including a modified abstract.

The search report should be used to compare your application, with the prior art, with regard to the novelty of your invention. On the basis of this, you may amend claims in your patent specification, to clearly distinguish your invention from an earlier disclosure. However, it is not possible to add information that you had not included when the application was first filed. At this point you must dispassionately consider the contents of the search report, since if there is a considerable body of prior art and you cannot reasonably amend your claims, then the application should be halted. Amendments must be made by filing duplicate copies of replacement pages. The search is limited in so far as it does not cover every item of information in the public domain and of course cannot cover unwritten disclosures. If this information comes to the examiner at any point thereafter and shows that your invention is not new and is 'obvious', then your specification must be modified accordingly.

During the 12 month period you may withdraw and refile the application, but you will loose the priority date. The extra time essentially restarts the clock, giving you more time to find a financial backer. However, there is a risk that during this period someone else will file on the same idea. This 12 month period after the first filing should be shrewdly used and a considerable amount of effort should be used in planning how you will use it. Patience should be exercised when deciding upon the filing date since a year is a very short time in which to prove the utility of your invention, especially in biotechnology research and development. Thus filing for a patent as soon as you have had your idea may be tactically counter productive, since the product may not be ready for introduction to potential licencees, i.e. they may perceive there to be too much risk.

☛ The third stage: early publication

Your application now becomes available for viewing by the public both at home (UK) and overseas, being published in the form in which it was originally filed. If you have submitted any amended claims these must arrive before the preparations for publication are begun, in order to be included in the published document. Otherwise they can be added at a later stage. The patent office will inform you approximately 10 weeks before publication of the completion of preparations to publish. All official correspondence between yourself (the applicant) and the Patent Office will also be made public and this should be borne in mind if you have information that you might not wish others to see.

The publication of your patent is indeed a watershed since anyone can now perform or make use of your invention in countries where you have not taken patent protection. It follows, therefore, that you must apply for patent protection in those countries before publication and usually before the priority date, since if you do not, any subsequent application could be refused on the basis that it was not new. Some countries (e.g. the US) offer a 12 month **grace period** post publication, whether this be in the patent or scientific literature. Some consider that this lack of uniformity gives the US an unfair advantage and the US first to invent rule (rather than the first to file) is now coming under scrutiny.

The published patent application may be used to show that any invention claimed in an application filed after the publication date is either not new or is obvious. This applies whosoever is the applicant. If your patent application is refused at the examination stage, then if you are post publication, you cannot then just abandon it and refile for the same invention, since the published application can be used to show that the invention is not new, i.e. is already in the public domain.

The monopoly right in the patent continues from the date of grant. If an infringement has been committed between publication and grant, damages may be payable once the patent has been granted. Thus, the early publishing of the application becomes very important, and when you are considering your application in light of the search report be sure that you publish any amendments or revisions of the claims. Publication may deter a potential infringer, but this is intuitively unlikely in the technology driven biotechnology and pharmaceutical industries. Obviously there is a reasoned risk to be taken, since: (a) publication does not necessarily lead to a granted patent; and (b) there is no monopoly until after the patent is granted.

There may be occasions where you wish to halt publication, which must be received in writing by the Patent Office before the completion of pre-publication preparations. For example, you may discover that your application is not 'new' from perusal of the search report. However, a modification may be possible such that you can introduce a new feature and make the invention 'new'. But, this will constitute an addition of new material, so you must now file a new application. Since the letters you send to the Patent Office will be open to public examination, it is not appropriate to disclose new information that you may think supports the additional application. Indeed, this constitutes a public disclosure and may preclude a further patent application.

☛ The fourth stage: the full examination

The fourth stage requires the Patents Form 10/77 and an examination fee, to be paid within six months of the date of the publication. Failure to do so may mean your application is considered to be withdrawn. During this examination, the patent examiner will determine if your application is in accord with the Patents Acts and in particular, will examine for the following criteria: (a) newness; (b) non-obviousness; (c) sufficient detail for the invention to be carried out by one 'skilled in the art'; (d) for clarity and consistency of the claims; and (e) for industrial applicability. As a result of this, the examiner will send a letter to inform you of the conclusion, which will indicate either compliance with the Patents Acts or not. You will then have a specified period of time in which to respond. The response should be either to revise the specification in accord with the examiner's comments or to present a reasoned argument as to why the appraisal requires further debate. This is an iterative procedure that will continue until the patent application is either accepted or refused. The path to a granted patent in now clear, providing all of the above steps have been completed within four and a half years of the priority date (or filing date if there is no priority date).

Other issues that may be pertinent to your application

If you cannot come to a reasoned position with your patent examiner (correspondence, telephone or informal interviews are possible) then the matter may be decided by a hearing, which will involve a senior member of the Patents Office. You or your patent agent will be expected to either write or attend the Patent

Office for the hearing. If the decision is not in your favour, an appeal can be made to the High Court.

Besides the amendments requested by the Patent Office you may make amendments of your own, between the date of issue of the search report (stage 2) and the issue of the first examiners letter (stage 4).

These amendments will be considered during the full stage 4 examination. A further set of your own amendments may be made in your reply to the first full stage examination letter. However, subsequent to this your own amendments may only be made by paying a fee and filing with the Patents Form 11/77. No changes can be made *at any time* if they constitute additional information.

Your patent may be withdrawn before grant by writing to the Patent Office; the effect is immediate and irreversible although some fees may be recoverable.

☛ Territorial issues

An important consideration is where to file the first application; a number of choices are available. The text to date has considered the UK application, which to summarise involves: (a) publication 18 months after filing, a compromise between the desire for secrecy and the public right to know; and (b) an international search that will be undertaken upon filing and which will establish novelty and be available within the year (Fig. 2.4).

The second option is that of a single application to the European Patent Office (EPO), in a procedure that is the same as for a UK patent application. This must be done through the UK Patent Office due to the national interest provisions. A search is carried out in The Hague and the examination takes place in Munich. All communications are in English. The fees for the EPO are much higher than for the corresponding UK patent, but of course the patent agent fees will be similar. The EPO route has the advantage that you will then also have a 'bundle' of national patents since one application gives protection in a number of different countries. Indeed, it was originally intended that the European route was cheaper than three independent national filings. If you had adopted the latter route, then in each case there would be a search and examination by the local patent office, often not in English. This would make the process very expensive. The protection afforded by a European patent will extend to all member countries, i.e. Austria, Belgium, Denmark, Eire, France, Germany, Greece, Italy, Liechtenstein, Luxembourg, Monaco, the Netherlands, Portugal, Spain, Sweden, Switzerland and the United Kingdom.

The third option is an international filing via the Patent Cooperation Treaty (PCT). Again, this must go through the UK patent office. This is as close as one can get to a world patent and is valid in any of the 74 member states, as of 1 January 1995. This route represents the majority of world patent activity and for applications originating in the UK, is administered by the UK Patent Office. The search that pertains to the application is subsequently published by the World

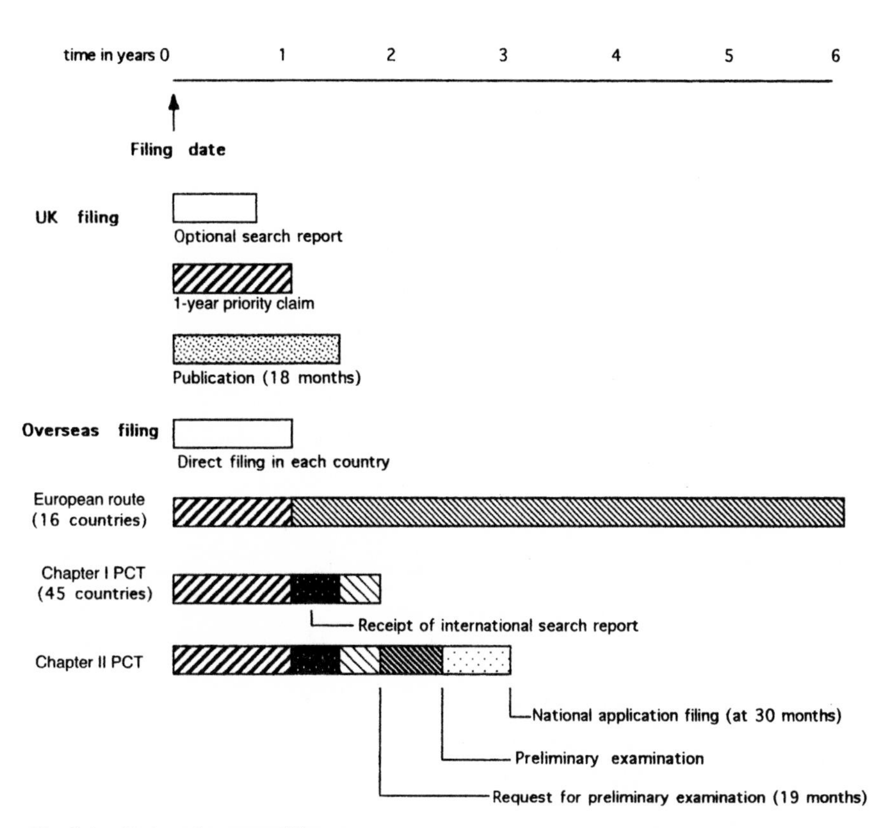

Fig. 2.4. Options for patent filing.

Intellectual Property Organisation (WIPO), which is based in Geneva. The advantage of a PCT is that the bulk of the costs do not need to be paid for two and a half years after the priority date. This means that you have a long time to assess the commercial potential of your patent and to establish a strategy for its exploitation. Indeed, by this time you should have found some potential customers and enlisted some help with the patent costs! The PCT has inbuilt flexibility and one of two routes can be taken. (1) In a Chapter I PCT, the application enters a national phase within 20 months of the priority date; you must however, select and file national applications. (2) A second route is the Chapter II PCT whereby upon receipt of (the same) international search report you may request a preliminary examination (19 months after the priority date). The resultant preliminary examination report (issued at 28 months) will allow a much better assessment of the strength of your patent; if you have a high value product (e.g. a new pharmaceutical) or an application that you consider to be weak, then this route may be appropriate. In a Chapter II PCT you must file the national applications within 30 months of the original priority date.

☛ The Cost of Patenting

There are two principle areas in which a patent will cost either you or your employer a significant amount of money. These are: (a) the fees to the patent office that are payable at each stage of the application; and (b) the fees to the patent agent. These are (approximate figures); £1500 for the initial application that gives 12 months protection, £5000–20 000 for the foreign filings 12 months after the priority date, £20 000–50 000 for the prosecution of the patent in a number of countries and £3000–20 000 for maintenance fees on every patent, every year post grant. Patent agents charge on an hourly basis and you are advised to check fees beforehand, especially if you are not using your technology transfer office.

The costs of acquiring a patent are very high and it is only of value if it gives you or your organisation a competitive advantage, which in financial terms exceeds the cost of obtaining and maintaining the patent. In many instances, several patents may be required for complete protection, thus multiplying the cost.

☛ Expectations from a patent

Patenting is very expensive, especially if you extend the application beyond the UK, which is almost essential for biotechnology based products. Thus the projected costs should be balanced against the true income potential. Much time and effort can be expended in the patenting process, resources that might be better spent upon selling your product in the market place. Many patented inventions never reach the market, often because the owners' expectations of the commercial potential exceeds the actual. It is often found that no one wishes to buy the rights to the patent, either because there was no market to start with or because the technology has been superceded. It is therefore essential that a thorough market survey is carried out before you consider patenting. This is not an easy task and there will always be a strong tempation to gloss over the relevant issues in the mistaken belief that it is expedient to do so.

There are a few negative aspects to patenting that should be considered. Patenting may not be the best way to protect your invention and other cheaper and simpler methods may be more suitable. Many inventions may not qualify on the basis of the lack of an inventive step, but this does not mean that they cannot be protected. The patent may not necessarily be yours to own for its entire life. It can be assigned to a third party or challenged. If the latter occurs then some or all of the patent can be lost. The patent will not protect your interests if you cannot afford to enforce it and, furthermore, there is no official route by which you can obtain finance in order to fight a patent battle. IP insurance may help but in reality your insurers will only let you enter the fray if they think you will win. Finally, the purpose of your patent is not to increase your personal wealth. It is to make the technology available for commercial exploitation and as such you must make the patent 'work'. Failure to accept reasonable offers may result in the grant of a compulsory licence.

The better the patent, the greater will be the worth and the more likely it is that it will be respected and/or challenged. Similarly, there may be a dominating patent in the field, perhaps to a generic technology, but one that might preclude utilisation of the patent. For example, any commercial developments that utilise the polymerase chain reaction will also require a licence to use this technique. In such cases your patent can be used as a bargaining tool to 'cross licence' and hence bring your idea to commercial reality.

☛ Significant differences in the US patent system

Unlike Europe, the US operates a 'first to invent' patent system and as such there are some important points to consider. In the US, the priority date is defined for US inventors by what they can prove has actually happened in the laboratory, e.g. by the use of notarised notebooks. For non-US inventors, the priority date is either the date of first filing or the date of introduction of the information into the US. It has been felt by many that this has given US inventors an unfair advantage, especially considering the importance of the commercial rights to exploit the technologies in the US. This situation has now been rectified and from 8 December 1995 it is possible to establish a date of invention for inventive activity in any WTO participating country. Previously, if a non-US inventor made an invention, patented and published without entering the US, then this meant that there was not a date of invention for use by the US Patent Office and the WIPO priority date would have been used. However, this has been a highly contentious issue and the interpretation of the US Patent Office rules are still subject to considerable debate, which often depend upon the circumstances of the individual case.

In biotechnology, the incidence of co-pending patent applications for the same invention is probably considerably higher than in other fields, and as such it is often necessary to make judgements on either the priority date or the date of first invention. Hence proof of invention as a criterion for the priority date in the US could give a temporal advantage over the first to file system as used outside the US. Thus in the US, rigorous notebook technique is essential, as is corroboration of laboratory work by third parties (non-inventors). The recent legislative changes regarding US acceptance of 'first to invent' in non-US WTO countries leads to a strong recommendation for you to adopt similar rigorous procedures, since given the frequent need to exploit biotechnology inventions in the US, a sound patent position will be essential.

There are two caveats to the above arguments that temper their effect; first many large companies, especially the pharmaceutical industry wish to have worldwide protection of an invention and already operate on a first to file basis and second, the patent office gives consideration to the uncertainty generated under the 'first to invent'; system, making the argument that one cannot prove **conception** unless all experiments have been performed, which would be in the patent application anyway.

One other contentious issue relates to the ability to identify all relevant patent applications, which can be very difficult for US patents since these are only published after being granted. Since US law allows for the filing of a **continuation in part** (CIP) or divisional application, the patent may not issue for several years making it extremely difficult to ascertain whether an infringement will be committed. This ruling is clearly to the advantage of US applicants since, contrary to those outside the US, they will have access to non-US pre-grant filings. However, in biotechnology, the applications are usually also lodged in other countries and these will be published 18 months after the priority date. If you do wish to investigate the patent position of a competitor in the US, you should request a search of the World Patent database from which you will be able to obtain priority details in the US. Thus, even though the specification and claims in the US application may be unavailable, this will help you to trace the corresponding non-US patents and patent applications. Furthermore, by following the progress of the corresponding applications through the EPO for example, you may gain an indication of the probability of the patent being granted in the US.

☛ Contentious issues

There is a high incidence of third party opposition in biotechnology patenting as one might expect from such a fragmented field where competition is extremely intense. The effects of litigation are both uncertain and expensive and legal proceedings should be avoided if at all possible. In addition, patent opposition and appeal procedures delay the processing of patent applications, which results in prolongation of commercial uncertainty. One solution lies in cross licensing, whereby both parties grant each other the rights to technology.

Given the nature of the biotechnology industry may be difficult to identify the inventive step. With the increased pressure to patent, some consider the threshold for the inventive step to be reducing, thus contributing to the plethora of patents pending. If you do intend to proceed with a patent, the advice of a patent agent who is experienced in the relevant field may well prove to have been an invaluable investment.

☞ *Protecting proteins*

The following example from US patent law will hopefully convey the necessity for the advice of a patent lawyer, preferably one skilled in biotechnology. The path to a granted patent is ultimately a judgement made by the Patent Office and there are often only a few precedents to follow; hence the requirement for a patent agent to skillfully argue your case.

Recombinant proteins *per se* are generally unpatentable if they are identical to a natural protein, with the definition of identical being that the two have immunological reactivity. To those in the field this definition is obviously inadequate, since many sera can cross react with related proteins or react with part of a protein, e.g. in a fusion protein. In these cases, the proteins are not identical. Furthermore, the whole point of producing a recombinant protein is to provide a cheaper alternative to proteins purified from natural sources, and thus mimicking a natural protein in activity is the primary objective. It is ironic that in order to patent such a protein you may need to demonstrate that it is substantially different in activity to that of the natural product. The distinction between a product and a process patent claim has been discussed already and many patent applications that involve recombinant proteins fall between the two schools, as a **product by process** (PBP) claim. However, many recombinant proteins show differences to the natural products in terms of activity, perhaps by being fusion proteins, truncated version, having amino acid substitutions or being altered in post translational modification. However, if the protein shows significantly different/better biological properties or indeed, the absence of contaminants, the patent examiner may allow the grant of the patent. There are now a few cases in the literature to which your patent agent will refer, so before launching into a costly patent in this area, you should take advice.

☞ *Moral issues*

The two major debates in biotechnology concern the patentability of: (a) life forms such as transgenic animals; and (b) DNA sequences without apparent function. The key point to remember is that standards of morality vary with nationality, time and individual perceptions. What may be 'politically incorrect' today may become acceptable within the lifetime of the patent and hence denial of a right today may be denying a valuable right 10 years hence. Even more unacceptable is that the right may be allowable in some countries but not in others, thus conferring competitive advantage to the country that purports the lowest moral standards, relatively speaking. Complex issues indeed!

Animal experiments must be carried out under a licence that has required ethical approval in the relevant country. Since these experiments must have been carried out by the time the patent is submitted, it follows that an ethical committee has approved the morality of the experiment. One should not confuse the issue of patenting (as an instrument of a technological monopoly) with the morality of animal experimentation; for example, the objections placed before the EPO with regard to the Harvard Oncomouse (a mouse which is pre-disposed to cancer by virtue of expressing the Ha-*ras* oncogene) concerned the ethics of interference with the genetic background and biological questions relating to the possible loss of genetic diversity. Despite originally reserving an open position on the generality of animal patents, the EPO balanced the risk to the environment and test animal against the advantage to the human condition, upon advice from the EPO

Technical Board of Appeal. The Harvard case is the only example of a granted European animal patent and to date, seventeen separate **oppositions** have been filed, each often involving many individuals, with the predominant theme of opposition being on the basis of contravention of **European Patent Convention** (EPC) Article 53a, i.e. *ordre public.*

One should therefore consider the three levels at which objections could be raised, those of fundamental ethics, public and scientific policy. With regard to fundamental ethics we must ask if patent law is the appropriate medium by which to discuss morality, rather than the other areas of law that are more amenable to shaping of societal behaviour patterns. However, since a patent is granted by a public body then perhaps there should be regard to issues that concern the public. The following are excluded from European patenting: 'Methods for treatment of the human or animal body by surgery or therapy and diagnostic methods practised on the human or animal body', primarily on the basis of a lack of industrial utility (Article 52–4). Also excluded are inventions that are contrary to morality or '*ordre public*', providing that the moral exemption is not due to legal prohibition on some or all of the states which are contracted to the EPC (Article 53a). However, as time proceeds it is increasingly likely that the national offices and European Patent Office will make judgements on moral issues, giving due regard as to whether the general public would, at the time, find the invention morally unacceptable.

Public policy must address whether the proposed invention is likely to cause harm to industry or the community, but whilst these issues are very real they are often expressed in an emotive manner with cogent argument *in absentia.* In particular, patenting of life forms is considered by some to lead to a decrease in genetic diversity, to prevent access to biotechnology for the Third World and to generally have a cost-benefit ratio that does not favour social advantage.

In terms of scientific ethics, patenting is considered by some to inhibit research and delay the publication and dissemination of information to the scientific community. Again, these thoughts are not borne out in reality since: (a) patents do not inhibit pure research; and (b) public disclosure may take place immediately post patent filing. In the former case, a patent infringement cannot be brought by experimental use of an invention and in the latter, it is unlikely that the results will be ready for public disclosure before the patent application is ready, especially since results are often kept secret prior to publication in a scientific journal. At the most there will be only a modest delay in publication whilst patent formalities are completed and, set against the experimental timescale, this is usually a minor concern. However, UK and European patents cannot be granted for 'plant or animal varieties or essentially biological processes for the production of plants or animals' with the proviso that this 'does not apply to microbiological processes or the products thereof' (EPC Article 53b). There are several other contentious issues, one of which is what constitutes an 'animal variety'. Unfortunately the term aquires a different meaning when translated into French and German, further adding to the confusion. Similarly, whilst microbiological means manipulation of microorganisms to most scientists, a patent lawyer may interpret this literally as a genetic manipulation of biological material on a small scale and as such, since the construction of a transgenic animal requires such genetic manipulation, then it could be patentable under EPC Article 53b. Clearly the language in some of the EPC clauses could be clarified, since when the legislature was first ratified in 1977, transgenic technology was not available and so it could not acccount for such cases.

In addition, the European parliament has issued a statement regarding animal patents such that transgenic animals 'shall at all events be deemed incompatible with public order and are consequently unpatentable', although this is likely to have little effect upon the EPO which is a largely independent body. However, some European countries have strongly supported transgenic work, so there is little conformity regarding this issue amongst EPC contracting states. In February 1994, the European Union (EU) adopted a Common Position on a Directive that seeks to harmonise exclusions from patentablility for biotechnology related inventions. The text therein is still the subject of discussion, but there are a number of clauses that will affect significantly the European biotechnology industry. These include the aformentionned transgenic manipulation of animals in the cases 'without any substantial benefit to man or animal', the use of gene therapy which is apparently 'contrary to the dignity of man' and patents that relate to materials derived from the human body.

The use of trademarks in biotechnology

In the initial stages of business development, it is quite right and proper for considerable attention to be given to the protection of intellectual property via patenting. However, attention should be given to trademarks as a method for **product protection** and creating value. A trademark is any word, name, symbol, distinctive feature, device or any combination thereof, adopted and used by a manufacturer or merchant to identify its goods and distinguish them from those manufactured or sold by others. A trademark will not protect an invention but can help to develop an intrinsic value, together with goodwill and reputation. As long as **renewal fees** are paid, a trademark can be a perpetual asset that has a lifetime in excess of any other IP right, despite the presence of competitors. If you are the first to enter a market, then any first mover advantage you have gained can be potentiated if you have built up a trademark. Once established with a trademark you have a legal right to stop others using that mark in a way that takes advantage of the goodwill, i.e. you can stop a competitor from 'passing off' their own goods as imitations of your own. When considering your prospects for registration you must first ascertain whether the mark is of a registerable nature and whether it is available. With regard to the former, the prospective trademark must be distinctive but not descriptive, i.e. the use of adjectives would not normally be allowed.

Companies often develop a family of related trademarks, which can be a particularly powerful way of distinguishing your products from those of your competitors. Trademarks can become a valuable asset and as many biotechnology companies are entering the market, often with similar products, it will become increasingly important to generate distinguishing features. However, it is crucial to use and enforce the trade mark in order to protect the goodwill and value that is created from having it.

3 The first steps towards commercialisation of your technology

War is nothing but a continuation of politics with the admixture of other means.

Karl von Clausewicz

In *vom Krieg*, Book 8, Chapter 6, 1834

Decide what you want and know where you are going

In bringing your research to market it is crucial to know what you want, to decide where you are going and where you wish to be. That is, develop a firm, flexible, strategic plan. So what do you want? Some options are: (a) to make yourself so rich that you never have to work again; (b) to have a career change; (c) to have other related commercial interests; (d) to have a lively interest in seeing your work developed further and used for the benefit of society; or (e) to make money for your institute. No doubt there are other possibilities. Your chances of reaching your goal will be dependent on the technology and its exploitation. If you have a technology that is only suitable for licensing then you can expect a percentage of the fee (depending upon your arrangement with your institute) and/or of the royalties. The more you can segment the potential market for the technology or the more generic it is, the more likely that you will sell additional licences. Perhaps the best example of this is the Stanford licence for genetic engineering (genetic manipulation using plasmids), but this is a precedent that is unlikely to be repeated often. Licences will: (a) generate revenue; (b) be a useful publicity aid to increase the profile of your institute; and (c) will help form collaborations with industry. However, if you wish to raise the stakes in order to make a larger amount of money for either yourself or your institute, then an independent start up company is undoubtedly the way to go. In this case the technology must be amenable to a start up scenario, e.g. a new generic method for making recombinant proteins. In addition, you must ensure that you have adequate IP protection and are able to convince the investors of a significant rate of return. It is almost inevitable that you will have to raise a significant amount of cash at this point.

There is clearly a difference between the two approaches due to the apportionment of risk. Licensing is, in general, a less risky strategy when compared to a start-up company. The rate of business failure is very high (hence the risk) and to offset this investors are looking for a high return on their money. There is a direct relationship between risk and reward and you must first assess the technology and consider how much time you are prepared to put into its commercial development. This may well alter your perspective of 'how much' money you are likely to get out of your technology.

It is possible to vastly overvalue both the utility of the technology in hand and the size of the research contribution to the final product. It is hoped that some of the considerations in later chapters will help you be realistic about the true value of the technology and thus gain credibility with your commercial partners. Whilst it is true that the product would not exist without the research contribution, similar arguments apply to the development, manufacturing and marketing processes. As a rough guide, the contribution of research cost to development cost is in the ratio of 1:9. Furthermore, the ratio of R & D to other business functions such as marketing is also about 1:9. Given the contributions of the other processes, you can expect the research to contribute around 1% to the final product. As researchers we only see what we do; it is wise to consider what contributions will be required from others.

The psychology of the commercial world

Many interactions between academics and those involved in commercial activities can be fraught affairs, beset with mutual distrust. A typical scenario would involve an academic entering the room thinking that he/she is going to be exploited. The negotiation position adopted can thus be defensive, intransigent and financially outside the bounds of commercial reality. The commercial contingent may feel that as soon as they appear, so do pound signs in front of your eyes. Many a good project has floundered due to suspicion and a lack of communication, long before the negotiating table. Many commercial entities actively seek new technology, but the chances of them approaching every academic on a speculative basis are low. The onus is on the originator of the technology to make industry aware of what is available. Increasingly, this function is performed by a technology transfer unit within the host institution but, depending upon their competence and experience, the alternative of direct marketing must be explored.

The first priority, therefore, is to ensure that you are in the right frame of mind for the negotiations in which you are about to participate. There is little point entering the room if this is not so, since you are likely to be wasting everyones time, including that which could be used on your own valuable research. No one will either know or understand the technology as well as you do and this makes you an integral part of the negotiating team. However, successfully commercialising a project requires many different skills and whilst your research skills are both necessary and indispensable, they will in general be insufficient on their own. In addition, your commercial colleagues are unlikely to have an emotional attachment to any particular technology. All will be viewed within the context of their overall business and the profits that could be made. The very nature ofbeing on the edge of knowledge often draws one inescapably close to experiments, concepts and results. The temptation and reality may lead you to regard the technology as a possession, which in turn brings out your protective instincts. It is often hard to see the cumulative work of possibly several man-years released to another, especially when you are unsure of the recipient's ability to look after it or if you feel you may have been exploited. These feelings are understandable and it is a position that will retain much sympathy amongst your academic colleagues. However, it is better to prepare your technology for release into the academic or commercial community and to retain sufficient influence to guide it through the development process into a technology much larger, much more profound and above all, a technology that is important to society. You will then be in a position to drive the project forward from concept to product and are less likely to feel as though you are loosing influence over your technology. You can enjoy seeing it develop into something much larger and more important than a publication in a scientific journal.

You should also consider the above philosophy in the context of your own motivational spectrum. Are you:

1 The pure academic who will have nothing whatsoever to do with applied research or commercialisation of the products of their research? Government policy is now forcing many to reconsider their views, as soft funding (i.e. grants from central institutions) becomes harder to obtain. No one is saying that you should not continue with 'blue sky' research; the issue relates to who should pay for it. Why not use the 'profit' from a commercial grant to fund your new ideas?
2 The academic who would commercialise if they could but does not know how to? Many would like to have more contact with industry and would like to see their research applied to new products. This is seen not to the exclusion of the normal research programmes, but as a complement. The problem faced by many is that they do not know how to go about it and there is a dearth of readily accessible dispassionate advice.
3 The 'commercial' academic? A small minority have already moved towards commercial exploitation, in parallel with their pre-existing programmes. This may be via set up of a company based on a particular idea and managed at arms length or with a more formal involvement. Titles such as 'scientific director' are often employed. These businesses have been set up with anything from garages and shoestring budgets to full scale venture financing and new facilities. Alternatively, such academics may be involved as consultants to particular commercial concerns. This is often in the form of a lump sum retainer in conjunction with an hourly rate.

4 The applied researcher? Traditionally the home of the 'new' universities, applied research is coming increasingly to the fore as projects of a developmental nature are undertaken within university departments. This gives the industrial concern access to: (a) the academic culture for highly competitive rates compared to in house fully allocated costs; and (b) the advantages of a creative free thinking academic environment.

Interpersonal skills

Looking around the academic community, the most well-known and popular scientists are all excellent communicators with the public. The ability to take complex scientific information and reorganise it into a form manageable for the general public is an enviable art. When selling your technology to the business community you must communicate your science in a clear, concise, comprehensive and objective manner. A woolly, negative, ineloquent monologue is not going to get the technology out into the market place, and you will be relying on the intellectual abilities of your client to realise the value of the technology. You will often be dealing with specialists in other fields. Thus you must be prepared to: (a) relate your technology in layman's terms; (b) Clearly state what the technology can do and what benefits it can bring to the customer (i.e. you need to do your homework on the potential customer); and (c) be positive at all times. Do not overstate your case or what the technology can do, but above all, do not say anything negative, unless of course, it relates to competing technologies. If in doubt, say nothing.

Much of the disdain between academic research and the commercial world derives from insecurity, i.e. the lack of comfortable reassurance. This in part reflects insufficient training and inexperience in each others' spheres, with the academic feeling out of depth with commercial procedures and the commercial negotiator feeling insecure in the laboratory. When you enter a room for commercial negotiations, you have entered the commercial world and your performance will depend upon one key issue, that of preparation.

Preparing for negotiations

Negotiation is part of life. Whether this is over extra staff, with grant giving authorities, extra laboratory space and extra resources, with the head of department or over who is going to do what experiment. With growing social awareness and a blurring of distinctions between environments, it is likely that you will become involved in more and more decisions as approaches to management change from hierarchical dictatorial styles to those of team based decision making. There are, needless to say, a spectrum of methods for negotiation:

1. Uncompromising. A battle of wills, take a position and do not move, hold out for a long time. Win at all costs. This could leave an embittered other side, with damaged relationships before the project or licence has been started or before the finance has been raised.
2. Compromising. Concessionary, 'giving in' to get an agreement signed. Avoidance of personal conflict. Amicable results are rarely achieved. The compromising negotiator ends up feeling bitter and 'done'. Probably they will also be over committed, under compensated and legally compromised.
3. Issue based. Probably a mixture of the two previous categories but with some extra points that will help you avoid barter and will result in enhanced relationships and fair agreements.

The commercial side are often hardened negotiators and will try every trick in the book. Scientists are often of an compromising nature, since in a negotiation room they inevitably feel that the commercial side is right (lacking confidence or not wishing to show ignorance) or have not prepared their ground sufficiently and thus are not in a position to argue.

In searching for a negotiated agreement, you should seek three points: (a) to produce a fair agreement; (b) to make sure relationships are either left intact or are improved; and (c) to make the process both efficient and effective.

Kneejerk negotiations

Two sides negotiate from a position, which is based on company policy and what they are prepared either to give up or pay. The danger is that as discussions proceed so each side becomes entrenched in its position and feels as though it must be defended, at an ever increasing cost. Hence the position becomes identified with individual egos and any 'giving in' may be seen as a 'loss of face'. Such an approach also means that you generally start from an extreme position, deceiving the other negotiator regarding your real position and, during the course of reaching an agreement, being prepared to make small concessions. The process becomes more and more time consuming and difficult since you need to decide what to offer, what to reject, how much to concede, for each issue. Hence this can lead to delayed proceedings. Personal resentment comes into the equation, resulting in strained and often shattered relationships. Thus, taking a negotiating position of this nature is: (a) not likely to result in a fair agreement; (b) can damage rather than enhance relationships; and (c) will be protracted at the least. These issues will be compounded if several parties are involved. The alternative approach of compromise negotiation may lead to a quicker agreement, but it is unlikely to be either fair or to leave both sides feeling happy.

So what is the alternative? How can you enter negotiations from a position of strength? With good planning, the outcome can be decided before you enter the room, by concentrating upon three key points. Depending upon the nature of the meeting, you should give it a proper quantity of quality time. In the first instance, define the problem and identify the interests of each parties. Furthermore, identify regions of self-interest. In the second instance, do not allow any party to be entrenched in a position such that moving from it can damage egos. Thus be highly objective and focus on the issues, not upon the people. Third, adopt a mutual problem solving approach such as 'here is a problem that we both have, how can we solve it together? Try to generate a series of solutions to the problem and spend your time deciding as to what the best solution is (Fig. 3.1). There are two key issues here: (a) it is good to have an 'ideas' person as part of the team, a lateral thinker who will generate many novel options; and (b) prepare your 'options for mutual gain' thinking beforehand. It is hard to think of creative, sensible solutions under pressure or if time is limiting. You will appear impressive if you can apparently come up with innovative solutions *during* the meeting. The approach taken will be to discuss each of these main points and finally, to discuss approaches to dealing with people who do not see the need to play by these rules.

Focusing on the issues

How many times have you had a problem that you have attempted to solve and ended up with angry, embittered people who have taken your comments to heart? This can even happen without you doing anything, e.g. by common gossip and wilful misunderstanding for attempted political gain. This begins a downward spiral which originates with the misinterpretation of statements that apparently threaten fragile egos and the ensuing reaction reinforces inherent prejudice. Thus it is important to be sensitive to the feelings of those around you, digesting the respective positions and looking beneath them to discover the interests of the parties. What are the real issues behind the positional cloak? Once you have found them, concentrate your efforts onto these issues *in toto* and seek to reconcile the interests of both parties, by adopting a joint problem solving approach. This will increase the range of options open to you, since behind these positions there may be many more areas of interest than there are of conflict. Indeed, there could be several possible final positions or solutions that will satisfy a diversity of interests (Fig. 3.2).

The key to identification of motivating interests is to put yourself in the position of the other party. Ask why (perhaps directly) a particular position is being held and do so not from a position of requiring justification, but as a means to understand the purpose and emotions behind it. In addition, consider the forces and pressures that may be acting upon the other party. Perhaps even try to discuss these with them. For example, what budgetary authority does this other party have? Do they regularly invest in external projects? What is their general position regarding IP? In dealing with the other party, remember that the negotiation is not

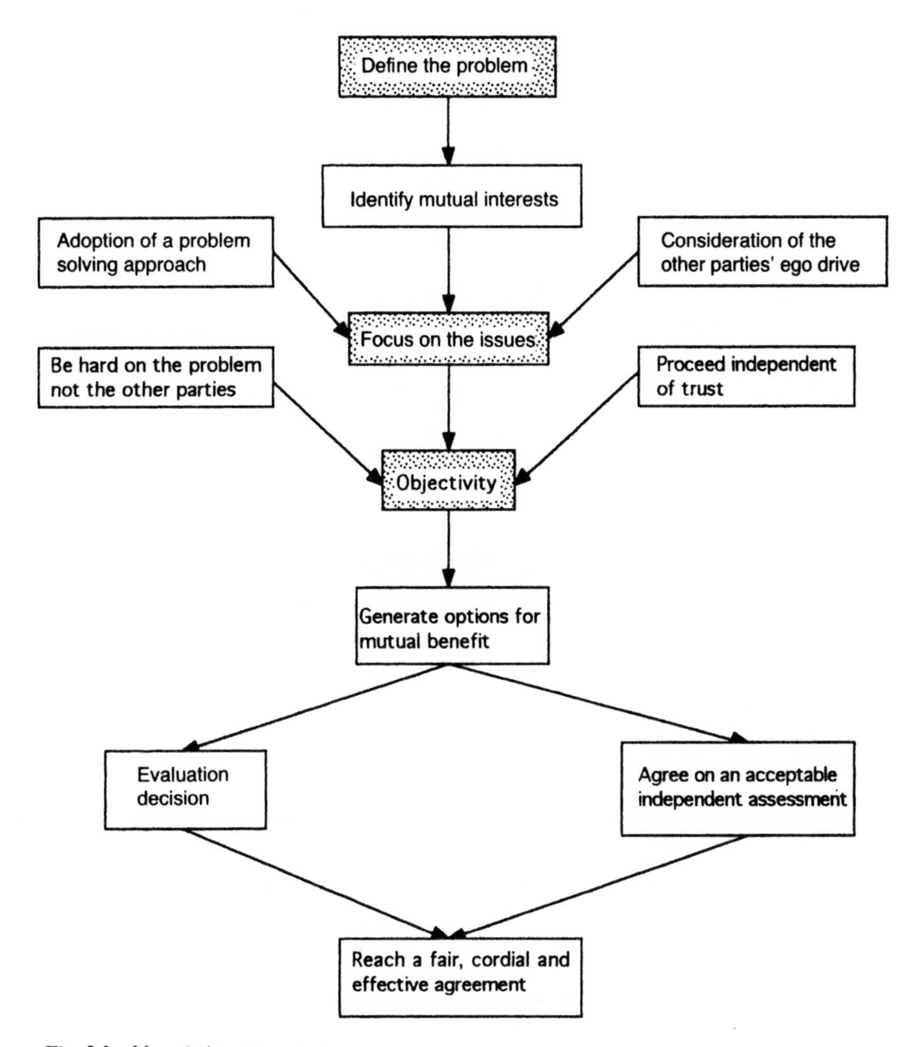

Fig. 3.1. Negotiating agreement.

just about money. You have to interact with human beings who have their own sensitivities and motivational spectrum. Most people have a need to satisfy their ego drive and your goal should not be to take this away, but to leave your fellow negotiators feeling better about themselves and also as if they have achieved something. This can be done subtly without compromising yourself (Fig. 3.3). Many times you will be picking up history, e.g. past grievances, beliefs, preconceived ideas. When these come out in conversations, do not reject or argue, accept them as valid points. Try to ensure that these concerns will be addressed on the way to the solution and as such, demonstrate that you understand the other sides

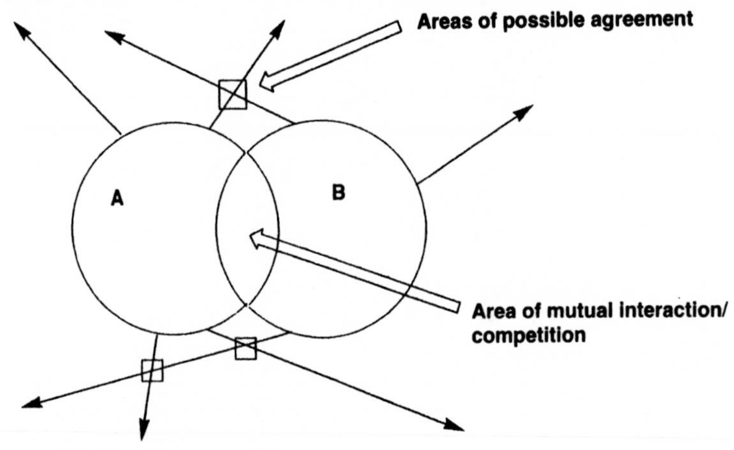

Fig. 3.2. Reaching agreement.

Problem	Solution
Blaming	Always counter productive. Elicits defensive reactions and counter attacks. Never apportion blame.
Perceptions	Be open, honest and clear. Do not trivialise others views, they may be strongly held for good reason.
The bolt hole	Reconcile the principles of the agreement with the self-image of both yourself and the other side. Use creative repackaging to avoid the appearance of one side giving in.
Participation	Involvement predisposes commitment. Make sure both sides feel that they have contributed. Give personal praise and credit to create an internal champion for an idea.
Fears	Avoid worst case scenarios derived from limited information, accentuated by your own fears and designed to discredit the other side. Creates a poor atmosphere if you are not careful.

Fig. 3.3. What to avoid and what to work towards.

interest. In this way, it is more likely that they will be of a receptive mind and prepared to listen to your problems and interests, which is the first step towards a jointly derived solution. What you must avoid is petty behaviour and points scoring. This may satisfy your own ego in the short term, but will be of negative value to the negotiating relationship, serving only to put everyone on edge. The line between this and good natured teasing is very fine and if you are not good at the latter it is better to play straight. In the negotiations you must be firm in statement of your interests. Indeed, be determined and definite yet be prepared to consider any new idea, since in this way you may create a position of mutual advantage. Consider two laboratories that have been offered a particularly precious 2 ml blood sample. Unable to decide who should have it, they divide it into two 1 ml portions and go their separate ways. As a result the first laboratory does not have enough lymphocytes for cell line isolation, but had discarded the erythrocytes, whilst the other does not have enough erythrocytes for oxygen uptake studies, yet discards the lymphocytes. Thus negotiations have left material (money) on the table and there has been a failure to reach the best possible solution, which would have been to work together on the blood fractionation so that both would have had enough material. So many negotiations fail for this reason and it is a fundamental error to focus upon positions rather than to negotiate towards mutual advantage. The principles that lie behind your interpersonal interactions in the laboratory are also of relevance to the negotiating room. In fact they are just the same, except the process is formalised somewhat and you are negotiating with people from different backgrounds.

We will now examine some options for lateral thinking; each of these are relevant to the search for a creative new solution that could lead to mutual advantage. First, do not assume that there is only one answer (an inevitable consequence of attempting to reach agreement via paring away two fervently held positions), when you should jointly be attempting to devise and hence select from a range of options. Second, aim for a mutual enhancement of $1 + 1 >> 2$, not $1 + 1 \leq 2$. Third, try to avoid the premature judgement scenario of which scientists are so fond. Great harm can be done by having someone in the room who is waiting to criticise any new idea the instant it is pronounced. Fourth, develop a sense of altruism. You may find that feeding someone else's self-interest also feeds your own.

In the negotiating process, first generate options, being scrupulous to exclude any evaluative steps. Arrange the session with a few relevant participants in a new informal environment. Mix-up the seating plan, be objective about the subject and record all ideas. Maybe then you can arrange a second meeting (or break the first) for people to collect their thoughts and turn to evaluation and decision. Such a process could be undertaken both internally prior to a formal negotiation or preferably with the potential licensee or financier. However, if you choose the latter, make sure all your staff are well-briefed as to what is confidential and what is not. During the decision process, you can then look for the agreement that gives you the maximum mutual advantage and the best grounds for co-operation. It is quite likely that you (as an academic representing a university department) will have different objectives to your commercial partners (Fig. 3.4). This should not be seen as mutual disinterest, but a way of avoiding conflict. In particular, you can emphasise positions of how the university technology may be used to increase the profile of the licensee in the local community, etc.

Objectivity is not necessarily a source of conflict. The latter arises from an inability to deal with differences, which in turn arises from fears and insecurities in individuals. Thus the more empathy you can engender, the more likely you are to reach a successful arrangement. My treatise in trying to understand cell biology has always been 'to think like a cell'. You cannot put human values into studies of how *cells* behave. These values are for how *humans* behave. Similarly, if you attempt to think like your negotiating partner, you will be able to understand their perspective and be in a better position to both help and reduce intransigence.

To do this, you must reserve judgement while you assess these new views since they will no doubt be as strongly held as your own. This is indeed not an easy task since each and everyone of us see what we want to see and tend to focus on a few facts which confirm our preconceptions. Thus side A will only see the merits of its own case and the faults of side B. There is an enormous difference between listening and digesting a point of view and agreeing with it. If this enables you to revise your own views, then this can only help the negotiation process.

A key to any negotiation lies in two-way communication. The latter is not always easy and is rarely done well, even if you have known someone for years (Fig. 3.5). The normal problems are compounded if you now have to communicate with those you do not know well and are in a formal negotiation, where there may be suspicion and perhaps even a degree of hostility. One of the most important points is to listen to the other side and indeed, the best negotiators are often the best listeners. Listening and rephrasing the other side's point shows that you have taken it in and acknowledged it. This is the most non-committal concession you can make. Another technique is to request clarification if there is any ambiguity. You must avoid polarization. If you do not give sufficient attention

University positions	Commercial positions
Ideological	Financially driven
Politically driven	Concerned with substance
Externally driven	Internally driven
Desire good relationships with industry	Focuses on the technology, no matter what the source
Case by case technology transfers	Relationships last the length of the agreement
Reputation, prestige	Regular licensee/licensor

Fig. 3.4. University and commercial positions.

One (or both) sides have given up and play to the gallery or to themselves
Spectators are deliberately polarised
Trying to catch out the other side (also deliberate)
Not listening (too busy preparing your next argument)
Wilful misinterpretation
Genuine misunderstanding Different languages

Fig. 3.5. Communication problems.

1. No matter what happens, do not react. If in doubt, take a short helicopter trip and look down at the situation. Alternatively, rise above it.
2. Continuously refocus on the issues and the substance under negotiation. Do not involve personalities and do not negotiate about positions.
3. Listen. Rephrase to acknowledge.
4. Acknowledge feelings and respect your opponents authority and competence.
5. Say yes as many times as possible without conceding.
6. Ask questions, do not make statements. Optimistically acknowledge differences between you as sources whereby you could work together to solve problems.
7. Create a cordial environment. Work together to solve problems, seek advice from your opponent and remember to ask 'Why' and 'Why not'?
8. If you spot a trick coming, expose it in a non-confrontational way.
9. Build your adversary a golden bridge (see Sun Tzu). Make sure your opponent leaves with his/her ego intact.
10. Proceed independent of trust.

Fig. 3.6. Ten key tips for negotiation.

to the other side, they will rephrase their arguments and repeat the process, with a consequent lack of progress. To promote a cordial atmosphere, rephrase their case in a positive manner and preferably better than they do. Then put your own case, being very firm on the points with which you disagree (Fig. 3.6).

☛ Too much communication

Be on your toes! If the atmosphere is hostile and you suspect wilful misinterpretation, some things are better left unsaid. Many negotiation problems come from unfamiliarity, both with the people involved and the technology. Thus attempt to build a working relationship as soon as possible. Discussing the children or holidays from cold can easily come across as trite and false. Try humour, or finding something that your negotiating partner is good at. Then ask for their advice on a particular issue. In finding this information you should try to gain it informally, i.e. wisely use the time before, after and during breaks in the meeting. During all interactions with negotiating partners always try to build the image of a problem that must be shared jointly in order to find an efficient path to a mutually acceptable agreement. Sometimes negotiations will get out of control and as the adrenaline rises, so do the emotions. Maybe there is a lot at stake and one side comes to the table in an emotionally charged state. Unless this is recognised and dealt with in an appropriate manner, these emotions will generate similar intensity on the other side, thus beginning a vicious circle of anger and fear. Clearly, whether it be on your side or the other, the key is to identify the source of the emotion. Is there some past history whereby revenge is sought? Has someone clouded the issue with events from another meeting? Bring these emotions out into the open, but do so without apportioning blame, i.e. focus on how you feel or how you might feel if you were the other side. However, as with all relationships it may be appropriate to let the emotional side release their emotions. As they do so, listen, do not react and make reassuring noises. Afterwards, they will feel better and believe it or not, nearly all of their missiles will have been fired. If there is emotion on your side, have a pre-meeting and work it out before you get into the negotiating room.

☛ Objectivity

At some point your negotiations may meet an impasse. For example, the absolute up-front fee for a licence will be a source of debate. Prolonged argument regarding such issues will be expensive and could lead to both inefficiency and damaged relationships. Thus to circumvent this, you must insist upon an objective measurement of the matter in hand. In the example given above, this is likely to be based upon precedent but depending upon circumstance, other measures will be possible, such as the use of prices of competitive products or an agreed accountancy measure. The precedent arguments makes you less vulnerable to legal attack but of course, many of you will not have access to the absolute value of such precedents. You are aiming for an agreement that is based upon solving the problem in hand and not one in which one party has won because of superior resolve. The principles of negotiating using objectives are as before; make sure that you are receptive to the standards to be used, make their identification a joint process and do not respond to bullying. Concentrate upon the issue of a fair agreement and although many tricks will be attempted, e.g. threats, the imposition of an immovable colossus of company policy or a submissive appeal for trust, it is not necessary to move from your position. Ask for both clarification of the argument and for it to be put in context of the objectives you have previously defined.

Problem	Solution
Preconditions for, or rejection of, negotiations	Investigate the reasons for not negotiating – status, internal politics? Use a third party to mediate. Re-establish the principles of the negotiation.
The suicide gamble	Forcing a major concession from which there is apparently no solution must be side stepped quickly, with a humorous or principled comment. Make sure you leave the other side an exit route.
Deliberate delays	Make it obvious that you know the delays are deliberate and search for another partner. Include objective deadlines.
Extraordinary demands	Relinquish the position to show that credibility is lost. Keep asking for rationalisation until the ridiculousness of the situation is obvious.
Increasing demands	Highly annoying, especially if closed boxes are reopened. Psychological submission on the part of your opponent, take a break and redefine the rules to suit yourself.
The threat	Quickly followed by the counter threat and destruction of the relationship. Do not respond, assume irrelevance, or suggest that it is unauthorised. Easy to turn to your advantage.
Personal attacks	Comments on your dress or IQ, keeping you waiting, not listening, etc. Bring it up specifically and deal with it head on.
Tom and Jerry	Nice guy/tough guy routine. Turn the question back to the nice guy or side step the issue.
Misrepresentation	Fact or fiction? Separate the issues and make your comments independent of trust. Adopt a policy of verification of all factual matters.
The power base	Make sure you establish the power base at the beginning. Do not allow the negotiations to falter because a higher authority needs to be consulted.

Fig. 3.7. Side steps and swerves.

These arguments are all very well, as a framework, but you will come across a negotiating team that has more experience, access to more relevant information or generally has an edge. How do you deal with such situations?

Difficult negotiations and different games with different rules

In many instances, you could be up against a side who are less than scrupulous. How can you counter their chicanery? An answer lies in following the aforementioned principles even more closely than before (Fig. 3.7). Most of the time, recognition of the tactic will be enough, since there is little point in firing the arrow if a shield can be put in that direction. However, be careful how you raise the issue, it could result in an open wound. As before, any comment must relate to the substance of the tactic and must not be directed personally. In this way, any umbrage or identification with personal ego is circumvented.

Experienced licensing lawyers are likely to give the academic a hard time, especially if you are not used to these situations. Most times you will have a legal advisor present but, if you don't, beware that: (a) an extreme position will be forcefully presented; (b) your technology will be to some extent trivialised (to reduce the price); and (c) your competence (in several areas) may be questioned. This will undoubtedly be a time to side step the problem, by accepting what has been said and treating it as one of several options.

Upon presenting your case and if you receive criticism, turn the table on the other side by asking what is specifically wrong with your proposal, i.e. do not react to the criticism by defending your position and digging in. Most of the time, this will have the other side on the back foot quickly. Now you are in a good position to turn the negotiations towards solving the problem together, which will be apparent to the other side as a concession, i.e. letting them restore balance and credibility.

At all times, it is wise to avoid the use of proclamations since the only thing you will generate is impasse. Thus reframe the discussion into questions that cannot be answered with yes or no. The advantage of a question is that it is free of criticism, leads to confrontation of a problem yet does not offer a point of attack. If a sufficient answer is not forthcoming, a period of silence will make the other side uncomfortable. They may well feel the need to justify themselves (to satisfy their own ego drive) and may give you extra information.

If there is a personal attack on yourself, let your opponents release their energies. One of their batteries will have been neutralised and you will have won the first confrontation. Then pause and refocus on the problem in hand, remembering that the objective is to satisfy your interests, not to hold your position at all costs.

The dangers of the bottom line approach

The 'bottom line' is the negotiating position beyond which you are not prepared to go. It is an extremely counterproductive position to take since: (a) you will probably have decided the bottom line amongst internal colleagues, i.e. those not involved in the negotiation, thus leading you to underestimate your case; and (b) it reduces your flexibility to respond to new information you may acquire during the course of the negotiations.

Instead, develop two or three final positions concerning the sort of agreement you would like to end up with, based on your predictions of the directions in which you think the discussion may go. It is good to have targets and something to work towards, but make sure there is some in-built flexibility.

Professional advice: what you should expect

Bringing your research to market will involve aquiring the advice of several non-scientific professional staff; solicitors (sometimes specialising in licensing), patent agents, accountants, business development managers (management of your business, market research and negotiating contracts). The issue here is that it is unlikely that these will be specialised in your technology or area of science. If you have been careful about your choice however, you should be able to find those who have some experience of biotechnology. This lack of specialisation means that attention must be paid to the extent of communication between yourself and your advisor. It is in your interest to make sure that the lines of communication are direct and efficient since you will be paying fees for their services. This section will give you enough information for you to know what to expect and what the procedures are. Similarly, in order to provide you the best service, the professional advisor will require pertinent pieces of information, presented in a particular way.

The role of the solicitor

It is strongly recommended that you seek the advice of a solicitor in all contractual matters. There are many that are skilled in licensing and company start up; the easiest source is to either ring up and speak to the practice directly, or search among your colleagues. In particular, you need to ask if there is any experience in biotechnology, in technology transfer, licensing or business start up, depending upon your needs. Your university/institute will undoubtedly have a practice that they regularly use or, you may seek advice from your university technology transfer office. The latter may involve you answering some probing questions. If you do not wish to use the office at this time (for various reasons), it may be a good idea to employ a solicitor whose name has been obtained from other sources.

Your solicitor is your sounding board and will be able to advise upon a number of issues, such as business start up, share allocations, legal documentation surrounding venture capital finance and licensing arrangements. If you have discussed beforehand the sort of agreement you think is fair, you will be able to enter the negotiating room with confidence. If you can afford it, ask your solicitor to go along with you. Regarding fees, there are a wide variety of hourly rates that are payable to UK solicitors. However, £120 per hour would not be unreasonable for a solicitor who is experienced in the area of interest. At these rates, you will not be able to afford a lot of time, so what you use, use wisely. For example, try to shift the cost of drafting an agreement onto your opponent, so that your own solicitor only has to read and comment.

Your solicitor will not necessarily have a precise view of your technology and it will be worth your time to give a general technical overview so that the agreement does not start off in the wrong direction. It may well be that your technology will fit into a relatively standard licensing agreement, which will give you a basis from which to work. It would be wise to have your solicitor draft a standard letter of confidentiality.

Solicitors can be expensive if used on an hourly basis, especially if: (a) two sets of lawyers disagree and begin an argument; (b) there is a discussion of a point of law; (c) if research into a point of law is required; (d) if rewriting of previously drafted sections occurs; and (e) if there are matters of drafting style to be addressed. In all cases you should act as a mediator and continuously address the commercial issues and their consequences, always searching for the shortest route to the answer and addressing the germane points. Lawyers often spend time solving problems left to them by businessmen, since some are quick to litigate. Try to make sure that you have exhausted all political and commercial avenues before involving the lawyers, since the legal route to problem solving can be a very expensive one.

The patent agent

Patent agents help you protect your intellectual property and will be able to help in a wide range of issues including ownership of IP, valuation of IP, patenting, licensing and protection of confidential information. A list of suitable patent agents can be obtained from the Chartered Institute of Patent Agents. Unqualified persons may offer advice in this area, but they are not bound by the code of conduct of the above institute. In addition, a good patent agent will be able to help European and worldwide patenting when necessary. Most large companies will employ patent agents who become specialised in particular areas. Since your industrial partners will probably have access to such specialists, it is to your advantage to make sure you have an equivalent standard of protection. The patent agent will advise you on the process, will prepare the specification and claims, and guide you through the technicalities and objections which may be raised by the patent office whilst it is reviewing your application. All disclosures to a qualified Patent Agent are confidential and are regarded as a privileged communication. There will be a series of fees payable to the Patent Office, in order to cover the application procedure and also to the chartered patent agent. To reduce the costs a little, you could draft the specification and many patent agents might appreciate this. The specification section should be similar to a scientific paper, but with a full disclosure of materials and methods used, i.e. it must be completely and totally reproducible by an independent laboratory. Patents can be rejected (in the US) on the basis of insufficient disclosure in this section and you should not necessarily expect accurate drafting of scientific nuances by the patent agent. However, the patent agent will expect to modify this section with appropriate legal language and you will need to work together on this point. The patent agent is required for drafting the abstract and most importantly the claims, where it is crucial to neither understate nor overstate your case. Either can produce significant problems. In addition, the patent agent will be able to carry out database searches if required.

Accountants

The accountant will be a key ally in several areas of bringing your research to market and will be essential from the earliest stages. In the first instance, the accountant will be able to help with registration for VAT, PAYE and income tax, in addition to dealing with the tax returns. If you wish to save money and do this yourself, remember that your efforts must be certified by an accountant under some circumstances. Help can also be obtained in preparation with the financial side of the **business plan**, whether this be presentation to the bank, to a venture capitalist or for some other purpose. This is particularly important in preparation of your **cash flow** forecasts since you must make sure you allocate all legitimate costs and do not overestimate your potential income. The accountant will also help you with financial modelling, i.e. a sensitivity analysis of your accounts. The presentation to various sources will require the use of standard accounting policies, so an accountant should be used to review your methods. The overall accounts of the business plan should be prepared by yourself, since you that have to justify them to potential financiers. The accountant can be used as a sounding board for these projections, before presentation to financiers.

Most importantly, accountants will advise in tax planning, enabling you to obtain maximum benefit from the tax system. This will include tax benefits on monies used as your equity stake, deferment of capital gains tax, tax on sale of equity as capital gains and unearned income and minimisation of inheritance tax. A number of legitimate business expenses may be offset against tax.

Business development managers

As you develop your commercial ideas, it will become obvious from the earliest stages that you need other expertise, particularly in the areas of negotiation, technology transfer, marketing, product design, information technology, etc. This is the role of the professional business manager, who will either have expertise in all of these areas or will know someone who does. An expert in negotiation skills would help in your licensing and financing negotiations and would be able to help draft the Heads of Agreement (see p. 96) and the final licence. Such a person would be able to read the licence from a business point of view (rather than the legal one) and advise as to the workability of the licence. For example, can the technology be transferred in the time specified and to the quality required? The business manager will also help with the logistics of technology transfer and may even be able to help perform it, freeing valuable time.

The most important area in which you will need help and to which you have probably given little or no attention is the marketing of yourself and your technology. The business development manager will be able to advise upon this subject together with the structure of a licensing deal (up-front fees and percentage royalties), finding and negotiating with potential business partners, advising on your IP portfolio, identification of areas where you could act as a consultant and perhaps most importantly, could help in the structuring of a start-up business. Such aspects of business planning will include the short-, medium- and long-term aims, market analysis, structure of the organisation, recruitment and selection of suitable staff, the presentation of the product, geographical areas in which to sell the product and how to get it there, how to promote the product in a cost effective way, efficient manufacturing processes and the use of information technology to make your business an efficiently run enterprise.

This plethora of business advice can be obtained at very reasonable rates in Department of Trade and Industry schemes set up to help small businesses (the enterprise initiative and one stop shops). If you are not yet a small business, general advice (probably not specialising in biotechnology) can be obtained from a variety of local schemes, such as enterprise agencies. More specialist help in biotechnology can be obtained from independent business development managers.

Breaking down the communication barriers

Your professional advisors will probably have only a small amount of knowledge of your specific area. Your business and commercial ideas will be only one of many that they continually deal with and whilst you may care about it deeply, they will see it as just another project. The advice you get will be dispassionate and given the experience of these people in their professional areas, you should listen carefully. You have a right to get value for money and understand what is being done on your behalf, so do ask questions. It has been known for confident scientists to insist on a patent application against the better judgement of their patent agent, only to find that they were right and that the fees have been wasted. Above all, maintain the lines of communication and realise that you may not know best in these diverse areas. There have been a number of examples of academics signing agreements with commercial concerns without either consulting their employer or seeking legal advice. This has led to severe underestimation of the value of the technology and more importantly, restriction on future activities for the academics concerned. In addition, due to the woolly nature of many academic employment contracts, there are often very few avenues by which to recover the situation.

Consultants and how to use them in the small business

Any experiment is spent 95% in the preparation and 5% in the doing. If you analyse your day, the majority of it is taken up with essentially non-productive activities. Consider then, if you also had to run the entire institute as well. Politics, seminars, hosting visitors, grant administration, personnel issues, etc. would ensure that you had very little time for experiments. Thus, in attempting to do experiments as well, at best all you could concoct would be a DIY approach to institute management (not being trained in the necessary areas) that would probably lead to an unsatisfactory administration and a lot of disgruntled colleagues. Compound this then, with conception, design and building of a new laboratory, recruiting and training the staff, raising money (not just grants) and a host of other issues. The problems and challenges are enormous. Obviously, the way to achieve these goals is to engage the services of someone who is prepared to do this for you and may well have additional training and skills in these areas. There are numerous management consultants who can provide the time, know-how and skills necessary to help run a successful business. Remember that your time is precious. Perform a simple calculation to find your hourly rate and consider if it is more cost effective to employ someone else to perform these tasks. Management consultancy is used in two ways: (a) to study/analyse a given problem and to give you a reasoned solution; and (b) to install a new system, idea or solution into your environment in order to solve a problem, i.e. to make sure that the mechanics behind a potential solution are in place.

A good consultant will not pretend to know about your technology in depth and the specialist knowledge on offer is likely to be in organisational issues, strategy, tactics, management of the company culture and so on. The service is likely to involve several steps: (a) identification of the problem; (b) derivation of a solution(s); (c) evaluation of potential solution; and (d) implementation.

A consultant should be both impartial and totally independent of your company and hopefully, with an overview that will suggest what is best for your business, i.e. be able to deliver a solution external to internal political influence. The consultant will probably be prepared to sign specific secrecy agreements you may deem necessary and will then spend some time listening, asking questions and evaluating the problem. Since you are paying for this service, it is in your interest to be open, frank and honest. The consultant will then produce a report, with a preagreed format, either verbal or written. During derivation of the consultants contract, you should agree the time scale, including performance milestones and the remuneration. It is crucial to have a clause that allows for termination by either party at any time, otherwise the assignment could extend in an unrequited manner, whilst at the same time costing you a lot of money. Progress reports must be stipulated and monitored, with remuneration being dependent upon timely receipt and pre-agreed quality. If you cancel a meeting at short notice, you may well be charged for time lost (since the consultant may find it hard to fill that slot with revenue generating time), so insert an appropriate clause into the agreement. Cancellation in excess of 48 hours would be reasonable. As to the cost, you can either pay a fixed fee that is tied to achievement of specified goals or charge an hourly or daily rate. Minimally, this could be £15 per hour, rising to £2000 per day for a doyen of the pharmaceutical industry. As a general rule however, the fees for decent management consultancy should be less than those for the solicitor or accountant. Between £50 and £100 per hour is reasonable, though you should always negotiate to get the best arrangement for yourselves. In some instances, you may be able to arrange for the consultant to be incentivised, e.g. by allowing receipt of a percentage of the level of business brought in.

There are many sources of help for your small business and in general, if you have a business problem you will be able to find assistance. The number of specialists in biotechnology are increasing, but be aware that the majority of problems you face will be of a generic nature and have been seen many times before. If appropriate action is taken quickly or indeeed, if enough forward planning is undertaken, you should have little problem with the mechanistic aspects. Actually selling the product or service will present a different set of problems.

4 The difficult problem of valuation of intellectual property

So the industrious bees do hourly strive
To bring their loads of honey to the hive;
Their sordid owners always reap the gains,
And poorly recompense their toils and pains.

Mary Collier,
In *The Woman's labour*, 1739

Basic commercial issues

We will now assume that the intellectual property (IP) you have generated has been appropriately identified, protected via a patent and packaged in a saleable format. This IP must now be realized in financial terms both to yourself, your institution and to the licensee(s). This requires a systematic quantitation of the value of the IP in question, decided not in isolation, but in comparison to similar technologies and as part of a business enterprise. The word 'value' is chosen carefully and should be distinguised from, for example, price and cost. The value of your IP should be considerably larger than either of these, since it also reflects the potential future benefits from the use of your technology as part of a product or service. Remember, the initial IP is usually only a small part of this product or service. In determination of the value of your IP, the next step is to consider a schedule for payment. This is often expressed as a royalty, i.e. an amount of money transferred in proportion to the contribution of the IP to the final product or service. Royalties are often expressed either as a percentage of product sales in monetary terms or of the total number of units sold. They can be paid with a variety of timings and very often this may include an 'upfront' payment, paid in advance of sales.

Valuation of IP requires an understanding of the time value of money and the difference between working capital, tangible assets and intangible assets. In assessing value it must be remembered that:

1 The royalty must reflect a fair rate of return for both the licensee and the licensor.
2 Due to the embryonic nature of much of the biotechnology IP and the rapid pace

of change in this field, there are considerable risks associated with commercialisation. This means that any quantification of the value of IP is likely to be subjective. Thus, is it most important to include sensitivity considerations in your derivation of a fair royalty rate.

3 By definition, your IP will be totally new and there will probably be very little information available concerning the pricing of comparable IP. In addition to the large risk element, prediction of the total value of sales that are likely to accrue will be extremely difficult. One might consider that by asking for a licence rather than outright purchase of the IP, the licensee is assuming some of the risk for product development and marketing. Hence royalites are most appropriately expressed as a proportion of total sales.

Besides deriving a fair value and hence royalty rate for the technology in question, consideration of the financial issues will help you see how the licensee or indeed potential investor will perceive you. An intelligent discussion of these issues using language that licensees or investors understand will add to your scientific credibility and help convince of your ability to transfer the technology.

The time value of money

Any investment, whether it be financial, personal or material represents a choice between alternatives. It is something we all face daily whether it be the choice between a high interest building society account and a personal equity plan or a choice between two different properties. At all times, sometimes subconsciously, we are weighing up the future benefits of presented opportunities and in some way putting a value on them. For example, some may prefer not to live on a busy road and attach value to a quiet rural setting. Others may be indifferent or indeed not prefer 'isolation'. In choosing our employment we often attach value to the location of the work place relative to home. Some may prefer not to travel and will compromise on the location of their residence. In these examples of course, we are discussing the concept of opportunity cost, which is the loss of revenue/benefit from taking one alternative over the next best alternative. So too with a financial investment whereby the potential licensee/investor will be faced with a number of alternatives to an investment in your IP. This may be a competing technology, a competing project in a similar or different field or indeed, a different business. Either way, the investor/licensee will be searching for the most profitable alternative. The most basic comparison is that with investment of a given sum of money in either your IP or a safer form of investment such as a bank. As we are all aware, a given amount of money now is worth more than the same amount of money at sometime in the future. This is because the money can be invested at an interest rate that will yield a financial return. The interest rate represents both a positive rate for expected inflation (the inflation premium) and

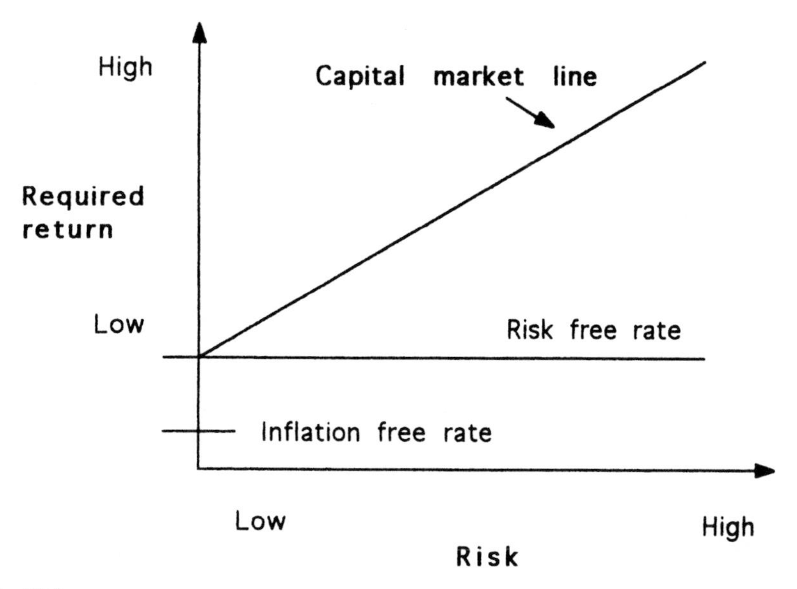

Fig. 4.1. Risk versus return.

an additional premium for time preference and risk. Time preference is an annualized ratio of the value placed on an item now and the value placed on the same item sometime in the future. Finally, the risk premium is the most important for us since it directly relates to the IP and can be influenced by the licensor. The risk premium relates to the probability of loss or the degree of in-built uncertainty. In financial markets there is considered to be a linear relationship between risk and return, i.e. the greater the risk the larger the return should be (Fig. 4.1). This is sometimes referred to as the capital market line. Even if the investment was risk free (a piece of IP guaranteed to come to market and reach the anticipated level of sales) there will still be a positive rate of interest that represents the premium for time preference plus inflation. Indeed, it would still be positive in times of zero inflation or coupling to index linked securities since a pure time premium would remain.

For the sake of argument we will consider the rate of expected inflation to be 5% and the premium for time preference and money market risk to be 3%. Hence the interest rate for a pure financial investment will be 8%. An investment of £1000 today will be worth £1080 (£1000 × 1.08) in one years time and £1166 (£1000 × $(1.08)^2$) in two years time. Hence the present value of £1166 to be received in two years time is £1000. The calculation can be reversed by *discounting*. That is, the value of today's £1000 in two years time will be £857 (£1000/$(1.08)^2$). In reality, the determination of interest rates is not easy and reflects the opportunity cost of the capital. The reader is referred to more detailed financial texts for a discussion of these issues. However, the important point for the owner of IP to realise is that the investor/licensee will be looking for a rate of return higher than that which could be obtained from another investment or deposit with a bank (8% interest). The required rate of return is often referred to as the **hurdle rate**. Hence it is important for you as the owner of IP to not only identify future uses, but attempt to quantify them. This will help potential investors see the monetary value of your technology relative to others (and for the most part you will not know what these are) and will help reduce perceived risk. The lower the risk profile the more likely you are to complete a successful transaction.

Investment appraisal

There are two commonly used methods for determination of the value of a project as an investment. These are the **net present value** (NPV) and **internal rate of return** (IRR). Both methods are in common use and will be used by: (a) investors, to assess the value of one investment relative to another; (b) management, to assess the value of one project relative to another; and (c) individuals, to determine if the returns from a venture are sufficient to justify the effort that is likely to be required. Both are calculated from a cash flow analysis and profit expectations, i.e. the today's value of those monies expected by a defined time in the future. The term '**present value**' refers to today's value of an amount to be received at some defined time in the future. The time dimension is accounted for by the calcuation of an IRR, the interest rate that equals the present value of the expected future amounts to be received. This means the interest rate at which one would ordinarily have to invest the money in order to match the proposed cash flows that will be achieved as a result of the investment.

The cash flow analysis should be the first financial statement you prepare when considering any new commercial venture or project. It is the key to your survival, being operationally more useful than other financial reports, such as the balance sheet and profit and loss account, since it gives an instant idea of how much cash you have in hand at any one time. A discussion of the other financial statements will appear in a later chapter. Ideally, the cash flow document should contain two parts, a highly detailed rolling one-year cash flow forecast starting from the present and a less detailed projection of cash flows in years two to five. These predictions can then be grouped in a cash flow diagram (Fig. 4.2). As indicated, a particular project requires an initial cash outlay (CF_0) and generates cash flows up to the end of year 5 (CF_1 to CF_5). The NPV is calculated by adding the initial cash outlay, which will usually be negative, to the present value of the anticipated cash flows in subsequent years. An interest rate must be defined, which could be chosen as the rate which you consider to be the minimum acceptable (a hurdle rate), e.g. the annual rate of inflation or that of your building society account.

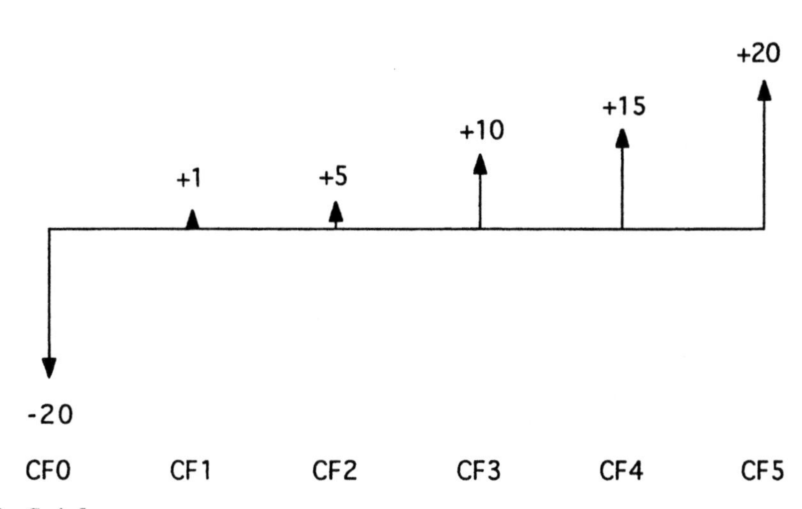

Fig. 4.2. Cash flows.

Managers and investors will all have different hurdle rates that will depend upon other projects in the business, investment criteria, etc. If the NPV is positive, then the financial value of the investment will increase and thus the investment is attractive. A NPV of 0 represents indifference and if negative, then there is a decrease in the investors assets and hence the investment is unattractive. In the example chosen the NPV is +22 using an interest rate of 5% and so this may represent a financially attractive proposition. The figure so obtained must be then compared to similar numbers derived from the other projects under consideration. The calculations for IRR are much more complex, being an iterative calculation to determine the interest rate at which the discounted future cash flows equal the initial investment, i.e. the interest rate at which the NPV is zero. As before, a positive IRR indicates a financially attractive investment. Depending upon the investment criteria this may or may not be attractive to the potential investor, but anything between 20% and 45% could be acceptable. In the example chosen, the IRR is 28%. If cash flows change sign more than once, then IRR calculations will give more than one answer, which makes them less useful. Furthermore, IRR assumes that monies received in the early years is reinvested at the same rate, which is usually untrue and results in a distortion of the real situation. The use of the NPV corrects this point since using this technique the money is reinvested at the **discount rate** used in the calculation of the NPV.

The value of IP in the context of the business

For the operation of any business, there are three essential accounting areas. This involves the management of: (a) working capital; (b) tangible assets; and (c) intangible assets. Working capital provides the financial means for carrying out the day to day running of the business. It pays wages, purchases raw materials and finances both current and new projects. The tangible assets are the structural items that could be sold and turned into cash relatively easily. This includes buildings, machinery and the employed labour force. The final category relates to the intangible assets such as IP (patents, copyrights, designs, trade secrets), which cannot be easily turned into cash. This means that if for any reason the business was wound up, IP is an illiquid asset.

Relative liquidity (ease of conversion into cash): working capital > tangible assets > intangible assets

However, one cannot underestimate the value of the intangible assets to the business. Working capital and tangible assets provide the 'mechanics', but the intangible assets drive the business by establishing markets, capturing customer loyalty, generating profit and indeed often determining the direction in which an industry moves. These criteria are particularly important in the relatively immature biotechnology industry where industry norms are still being established. Many new biotechnology companies have a relatively weak intellectual property base and are involved in a continuous search for technologies that will strengthen their positions. This creates considerable debate regarding future value and appropriate pricing. Indeed, many new biotechnology companies are based on anticipated benefits from patents applied for (high risk!) and often only one or two (sometimes none) of these. As previously discussed, the intellectual property base in biotechnology does not yet appear to have reached a *status quo* and there is much debate that concerns the disbursement of generic (enabling) technologies. This uncertainty increases risk relative to other industries, but of course the rewards (being increased in proportion to risk) can be enormous. A particular sector of investor is attracted to higher risk/higher reward investments in companies that are based upon new intellectual property, these being as their name implies, the venture capitalists.

Intangible assets are a fragile commodity and both come and go. This can be equated to replacement of a given player in a sports team. A 'star' player may well make the difference between winning and loosing, yet the tangible assets (e.g. the stadium) and working capital are likely to remain the same. In more mature industries we are accustomed to viewing business success in physical terms such as large well-equipped laboratories and a competent trained workforce, i.e. we equate profit to the efficient use of both capital and labour resources. However, this preoccupation underestimates the contribution of the intellectual property since it is the most effective way of demonstrating differentiation from everyone else. Thus when viewing your IP, try to examine it in terms of the contribution it could make to a given business. Try to anticipate the future markets and *value* to the licensee's business. Again this is of particular importance to the biotechnology industry since, whilst there are now just over 20 recombinant therapeutics on the market (and that this use is likely to increase by the turn of the century), there is still relative uncertainty (e.g. in comparison to chemical synthesis) relating to new products likely to arise from new IP. Hence, there is potential to 'hype' new IP, but beware that you are also increasing the risk and maybe creating false expectations.

Valuing the contribution of your IP to a potential licensee's business may seem like a difficult task. However, given some time and access to relevant sources of information it is often quite amazing what can be deduced.

Valuation principles – how to exploit your IP, when and with whom

The intellectual property to which you have access has value. This must of course be true, otherwise it could be published without any thought of protection. The next question, is to whom is it valuable? Exactly who is likely to want the technology and how much is it worth to them? For example, let us consider an ordinary piece of laboratory equipment such as an ultra-centrifuge; what is the value of this item? This of course depends on the beholder and one might envisage a different value according to: (a) the users; (b) another laboratory (who maybe do not have one!); (c) laboratory heads and/or the university/research institute; (d) replacement; (e) scrap; (f) liquidation; (g) part exchange; or (h) insurance. Thus, the first step is to list the routes by which your IP could be exploited and then to pick those with the highest value and best uses. As will probably now be obvious, this can be via a continuation of the development process and marketing the IP yourself, forming a joint venture with someone already in the market or one who wishes to enter it and/or licensing directly to third parties.

The value of your intellectual property is also affected by the 'speed of sale', which in most cases will be in the 'orderly exchange' category. This means that you have time to locate a buyer in the market and enter into a well-negotiated agreement. Alternatives might be a forced sale to an intermediary for 're-packaging' in which the revenues will be reduced due to the costs associated with the intermediary, or the worse case scenario of immediate disposal, say at the equivalent of an auction, which is likely to result in the lowest price. The role of the intermediary is likely to become more prevalent since many institutions with diverse research interests have now built a similarly diverse IP portfolio. It must be questioned whether they are properly qualified to market this technology and hence a brokerage or 'IP bucket shop' may be a viable route to commercialise certain IP. Entry barriers for such a venture are likely to be high, since credibility can only come from having access to a large IP portfolio.

The value of the IP should also be reflected in the effort expended in obtaining it, i.e. the cost of reproducing the technology at the time of appraisal. This includes all resources (man-hours, materials, overheads, etc.) that have been expended during the execution of the project. However, technology moves at a considerable rate, especially in the biological sciences. This means that it is easily superceded and the rate of depreciation is accelerated by the plethora of similar patents that follow closely behind.

How then to place a value on the IP? There is no doubt that standard accounting measures are unsatisfactory since they only serve to severely underestimate the value of the IP in question. Indeed, these arguments may be used as a counter measure by your opponent in negotiations. Beware of arguments that use: (a) Original (reproduction) cost for anything other than a starting point in analysing price trends; and (b) net book value (original cost minus depreciation). Ideally, you should attempt to quantify the future benefits and calculate the net present value of the IP, i.e. the sum of the expected future benefits discounted back to todays money. The discount rate will reflect the perceived riskiness of the venture. The future benefits will include income from all sources (licence fees, royalties), contract research, consultancy contracts, sales from use of the invention, etc. Estimation of economic life is a relatively difficult and inaccurate task for the biotechnology arena. There are several methods available from other fields, all of which are based upon historical data. These data are then applied to groups of similar assets whereby the historical experience is expected to provide a guide to the future. However, this approach requires that all of the necessary historical data, i.e. the complete life cycle of a similar asset, is available for incorporation into the calculation. For high technology IP there is often no history and an uncertain future, with high associated development costs. If there is a considerable risk of technological obsolescence then the royalty rate will be reduced correspondingly since the technology will have lower value to the business and there will be a lower expectation of future revenues.

There is a school of thought that believes it to be possible to value the IP outside of the particular business enterprise. In some cases this may be true, but in reality if you can estimate the value of the technology to the company in question, then not only are you in a stronger negotiation position but you have an opportunity to increase revenues. For example, in a recent case it was established that a particular company had lost all of its drugs in phase III clinical trials and had a very poor pipeline for short- to medium-term revenue generation. This was researched because one piece of technology within our remit involved the circumvention of a side effect of a particular drug and indeed, this contributed to the drug authorities not approving the phase III clinical trial. Obviously this piece of market intelligence could put your team in an excellent negotiating position. Indeed, the patent in question complemented their existing portfolio and provided one last brick in their patent wall.

A further consideration in the valuation of your IP is the cost of proving that it has a commercial future. This is not often appreciated by the academic who inevitably believes his/her technology to be pre-eminent. To prove industrial utility may require steps such as; production of a prototype, drawings, manufacturing specifications, testing, large scale batch cultures (for cell lines or bacteria), clinical trials or regulatory considerations. Indeed, some of these may require several iterations. Without this information the value of the patent to the market may be reduced and you should at all times build these considerations into the timing of your approach to the market. If you have reduced the risk to the potential client, you will have arguably increased the price. This is often a matter of a few more experiments, collection of clinical data, etc. and if this can be accomplished in a reasonable time frame, then you may be able to increase the value of the IP to the potential licensee.

The life of a piece of IP follows much the same path as the bacterial cell growth cycle. There is a lag phase as the IP becomes recognised, an exponential phase when it is taken up by a licensee, a maturity phase when it generates revenue and finally a decline phase when it is superceded by new technology. The exact path followed will vary for the industry and product, but we would expect these to be relatively short in biotechnology. The life of the product is linked to the value it has in the market place and the best pieces of IP in terms of longevity and value are those where a generic process finds many end uses. In this way, diversity of use shields a downturn in any individual area, thus reducing the risk. Most of the readership of this book will be familiar with the impact of the polymerase chain reaction (PCR). This is a particularly fine example of a generic technology where there are many 'end users' and hence there are likely to be many valuable end products derived as a result of its use. However, the economic life of such IP cannot exceed its market life and there are several external factors that could affect the economic life of the IP. Each of these should be borne in mind when considering the position of the IP in the overall market, since this will also affect value. These factors are: (a) the probability of a competitor designing around the IP (e.g. there are now at least eight DNA amplification systems in addition to PCR); (b) the probability of a challenge to the patent; (c) the potential impact of new legislation; (d) unanticipated high development costs that render the IP uneconomic; or (e) price escalation/loss of supply in an essential raw material.

The economics of royalty payments

The amount of royalty negotiated must reflect a fair rate of return on the value of the IP. There are good reasons for making sure this is approximately correct since overpricing could drive away any other potential licensees and underpricing could result in a serious loss of revenue. The goal is to ensure that the licensee attains recompense for use of an asset and that the licensee benefits from an asset which would otherwise be unattainable. However, IP is difficult to turn into cash (i.e. it is illiquid) and generally has poor versatility, in that it is often difficult to see how it might be redeployed elsewhere within the business.

In calculating royalty rates, which will usually be as a percentage of product invoice value (carefully defined), units of production or degree of use, the goal is first to identify the amount of net income that may be generated by the IP. As before, this is the present value of the future stream of total economic benefits that derive from ownership of the IP. In the derivation of such a figure you will also determine whether the monies will be increasing or decreasing, for how long they will be paid and the extent of the risk, expressed as the probability of obtaining a particular amount in any given year. The present value is dependent upon the economic climate and the value of the IP will be in proportion to the contribution to profits that it makes. The latter is difficult to calculate but if you do attempt this, remember to include the amount that would have otherwise been paid in licence fees.

Investors/licence negotiators may use a variety of calculations to calculate a royalty rate, but these are often fraught with rather gross assumptions in all but the simplest of cases, since reliable data is often difficult to obtain. The empirical methods fare little better, but will nevertheless be outlined below.

☛ Use of industry norms

This uses the figures that others may use for licensing IP in the same industry. In fast moving fields the value of the IP may fluctuate considerably but if using this method for a generic technology you should probably pitch at between 10 and 20%, reducing this for more specialist applications where the risk of failure is higher. If you have a strong IP position, then the eventual (negotiated) royalty rate could be around 8 or 9%. Weaker positions may result in a final royalty rate of between 1 and 5%. This method does not of course consider any requirement for complimentary assets, for investment risk, growth potential, current market size or profit expectations. Any mistakes made during previous royalty negotiations can be passed along and of course, this method cannot account for changing economic circumstance.

Some consideration should be given to the size of the potential market for the product; this can be achieved by desk/database research. It is useful to estimate the sizes of the world, European, UK, US and Japanese markets, together with the market share of your prospective client in each geographical area and the increase that it might be possible to achieve using your technology. This will enable a calculation of the absolute monetary value of the technology to the licencee and hence be a guide to: (a) the figure set for assignment of the IP in its entirety; (b) the amount of the up-front fee; and (c) the guaranteed minimum royalty payments.

☛ Return on sales

Again, when several pieces of IP contribute to product sales, it is extremely difficult to determine the contribution to profit of any single part of the IP. Also, it is almost impossible to provide a weighting for investment risk or the phase of the IP life cycle. Sometimes one hears the argument that particular products could not be sold if it were not for the contributions of the licensor's IP, which is usually an effort to increase the price. However, one should be referred to the arguments that relate to a fair rate of return for the IP in question. Nevertheless, this is a common method of negotiating royalty rates, often using a figure conjured from thin air, which then forms the basis for a series of risk allocation negotiations. However, beware of the figure that is pitched wrongly in the first instance.

☛ Return on research and development

R & D costs will rarely if ever reflect the true value of the IP in the market place and consideration of our previous discussion has emphasised the value of the IP in terms of future economic benefits. R & D costs will not reflect the costs and risks of obtaining income nor the amounts involved.

☛ The 25% rule

In this method royalties are calculated as 25% of gross profit before tax. In the initial negotiating gambit the figure should be set at 30% or higher. Thus the contribution of the licensees costs will not be considered, even though they will be negotiating as if they are considerable. Clearly then, IP that does not require great expense for the licensee in terms of marketing, human resource, etc. will be more profitable and the licensor should therefore receive a higher royalty. That is, if two products contribute the same gross profit and one requires a considerable selling expense, then the other will contribute more towards profit, thus arguing for a higher royalty.

Final considerations

All of the above methods are not totally satisfactory due to: (a) lack of consideration of investment risk and return; and (b) the absence of a weighting for the contribution of current assets.

To summarise:

1. Information relating to royalty rates for similar IP is difficult to obtain and reasonable comparisons are few.
2. Industry royalty rates have been often established for some time and may not reflect changing circumstance.
3. Royalty rates may have been negotiated with minimal consideration of business risks and rates of financial return.
4. The royalty rate chosen may reflect the inclusion of specific clauses into the licence, which are likely to be irrelevant to the technology in hand.
5. The competition may affect the magnitude of future cash flow.
6. Your patent protection may not be absolute since alternative designs/technologies could have become available.

As should now be obvious, there is not a clear answer to the question of 'how to calculate royalty rate?' It is a subject of intense negotiation and the advice herein is to collect as much information as you can on the markets, potential licensees, their competitors, alternative technologies, etc., in order to build up your negotiating position as much as possible. You will be entering the room with incomplete information, but of course, so will the potential licensee.

5 Developing your ideas

One of the greatest pains to human nature is the pain of a new idea.
Walter Bagehot
The 'Age of Discussion' in *Physics and Politics*, 1872

Introduction

This chapter will consider the issues of developing your technology further. This will involve marketing yourself and your product to potential licensees, financiers and/or technology transfer companies should you choose to use one. It will involve looking at how to find potential licensees, how to segment the market and what information is publicly available to you. The next section will look at the options for commercialisation and how to choose between them.

Consultancy and contract research

The 'top end' of the technology transfer business and therefore also that of taking your research to market are the business start up and licensing deals. There are however, two other areas whereby commercial links can provide revenue and more importantly, can begin to form effective relationships. These lie in the areas of: (a) contract research, whereby industry sponsors a definite piece of work for a given period of time, using given resources and with an expectation of a defined end result; and (b) the general marketing of consultancy services. The latter involves the wide range of academic expertise in the university that can be sold to the benefit of all sectors of industry. In marketing yourself and your technology, you should not be quick to discount these avenues. They may not appear as attractive as the 'start up' or a licensing deal, but they do begin to establish relationships and build trust, which is essential to the establishment of cordial and productive negotiations.

☛ Marketing yourself as a consultant

Industry uses a considerable amount of consultancy from academic institutions. This is often arranged directly between the academic and the industrial partner and usually suffers two major drawbacks; the pricing of the consultancy is generally too low and most importantly, there is rarely an attached work schedule. The scope of the project is thus left very wide and the inevitability of leakage of valuable university intellectual property is obvious. Whilst the individual concerned may be delighted with a few extra pounds in the pocket and a considerable boost to the ego from the 'consultant' title, the loss to the individual and the university could be considerable.

What constitutes a consultancy? Any money or resources (payment in kind) that you receive from a defined activity during the normal course of your contracted employment duties is a consultancy, not including of course, your basic salary. This will include lecture fees, retainers, written or oral advice and so on. Many will choose not to direct their activities through their universities or institutes, perhaps believing either that they will receive personally more money or perhaps they may even resent giving their employer a 'cut' (even though they have an employment contract). The unusually loose relationships in the employment contracts of many academics in UK universities allows some liberties to be taken. One of the fundamental areas to tighten up will be these employment contracts, otherwise the increased commercialisation will be non-sensical. If you decide not to use the university facility for management of consultancy (even though the taxpayer has probably funded much of your research and provided the means by which you obtained that expertise) then you should: (a) not use the university address/ headed paper in any correspondence or run the consultancy from your workplace (i.e. do it from home); and (b) take out appropriate personal indemnity insurance. You and you alone are responsible for the consultancy and the university will not be able to accept any responsibility for your activities if you do not use appropriate channels. Indeed, for the avoidance of doubt, you should perhaps write to the client and inform them of this point.

The alternative, i.e. that of processing the consultancy through the university system, points to an altogether preferable picture, in that you will probably be able to use the personal indemnity cover of the university and you will be entitled to full legal protection. Generally you should obtain permission from your Head of Department, pro-vice Chancellor, Dean of Faculty, etc. and in many cases you will be limited to less than 30 working days consultancy per year.

Similarly, from the clients point of view a central consultancy service has a number of advantages such as a professional corporate identity, a central point of contact and thus a quick response, and probably also a database whereby multidisciplinary teams can be assembled in order to solve larger problems.

Hopefully your university/institute will benefit both financially and politically from the consultancy you undertake. Remember, that you are an ambassador for your employer and as such you must ensure that the reputation of the same is enhanced.

Clearly, if you are acting as a consultant you must attempt to get as much money as you can, at a rate which reflects your true market value. For a professor of a department who is consulting on a large project in the pharmaceutical industry, the fees should be between £800 and £1500 per day. This figure will shrink dramatically as one goes down the hierarchy, ending up with senior postdoctoral fellows. A postdoctoral could expect to charge around £20–30 per hour for services rendered, plus expenses. Very often this is circumvented by inviting you to give a seminar, for which a fee will be paid and the prospect of employment may be alluded to. Fees for giving a seminar to a commercial organisation will be up to £300. It would be surprising if more junior employees had the depth and breadth of knowledge necessary for a serious consultancy contract. It is more likely that if at this stage in your career, you will have developed or know a specialist technique and will be able to transfer this technology on a one-off basis. However, it is critical that you: (a) check your employment contract to make sure that you are allowed to take on such consultancy; and (b) that you check with the relevant authorities in your institute to ensure that you are not compromising an intellectual property position. Whilst the money on the table might look a lot, you can be assured that it is little when compared to licence fees or royalties. The unwitting release of information to the wrong people could compromise many other issues. It may well be that they have additional contact and may be able to negotiate better terms than you would otherwise be able to do yourself. In addition, it is only right and proper that your employer be compensated for an extensive amount of time spent in consultancy, if indeed you are allowed to do it. After all, they are already paying for your time and if you employed someone to refit your kitchen in a week and they kept leaving the job to do work for someone else, it is likely that you would not be amused.

☛ Networking

Perhaps the most important part of the marketing process is a technique known as networking, whereby you build up a portfolio of personal contacts over time. Some exploit this technique to others disadvantage and err on the side of abberant business ethics by pretending that so and so (usually a senior person) has suggested they speak to you, etc. One also has to be careful about the contacts made since it is easy to slide into a 'network of disrepute' without being aware of it, especially in a field with which you are not familiar. Other less scrupulous networkers may use your good name in vein. This can be very irritating, difficult to prove and even more difficult to prevent. It is, however, inevitable that you will encounter those whose business ethics are less than acceptable. Thus, you are advised to build up a network slowly at first, making contacts via seminars, conferences, colleagues, etc. whilst at the same time being careful of the personalities that are allowed into the inner sanctum. As you move into the business sector it is much harder to assess credibility, since you cannot refer to someones published work and assess it in a quiet moment. The opinions of people within the network thus become paramount and it is at this point that the value of the network will become clear. It is advisable to keep a list of names, addresses, etc. on a suitable database programe for easy access, addition and comment. Make sure that you use the network to its full extent by calling people as necessary to seek advice or further contact names. Cold calling is rarely effective and the networking process will aid your personal marketing both academically and in the business world.

Contract research

Following closely upon consultancy contracts the next type of commercial arrangement you may be approached to perform is a particular piece of research, in return for compensation for your department and/or yourself. The reasons for doing so will include recognition of the academic/research expertise within your laboratory, reduced cost compared to the commercial laboratory and a lack of human resource/expertise in said laboratory. There are both advantages and disadvantages in carrying out such research that will come to light as you read the following section, which delineates tips for inclusion in the contract research agreement. The major advantage is that you will be forging a contact with industry, which can lead onto other relationships such as larger grant funded research, studentships, etc. The distinction between contract research and contract services is somewhat grey. If you are providing a service that does not require the development of new technology such as a diagnostic test or genetic screen, then this cannot be considered research. Contract research must involve an attempt to find an answer for a problem which does not currently have a solution. The problem will be of a highly applied and confidential nature that specifically relates to a product or process relevant to the commercial concern. Between this and the conventional research council grant can be a spectrum of relationships, one of which involves the government LINK scheme, which will be considered in 'Collaborative research schemes' (p. 82).

The nature of the contractual relationship in contract research

One of the first issues raised by many academics when faced with a contract research agreement is the issue of publication. For most, the need to publish regularly is the route to career progression and as such, it contrasts to other professions where ability is judged by other criteria. It is unreasonable for these genuine concerns not to be attended to in a forthright manner and indeed, maintenance or enhancement of UK publication rates are an integral part of demonstrating the strength of the research base. Many agreements will contain clauses that restrict publication and these should be carefully considered. After all, publications are necessary for your career, the careers of your staff and for raising additional research funding, being the pre-eminent justification of previous success. There are two reasons for wishing to delay publication. (1) The time required to consider whether or not to protect the IP via patenting. A decision should not take more that one month, but this of course will depend upon the company concerned. The issue of IP protection must be specifically addressed in your contract research agreement. (2) The competitive edge. It may well be that the sponsor will seek to maximise the time in which to consolidate its competitive advantage from your research project. In this case you will be a victim of your own success, since the more productive you have been the more likely they will wish to protect the information! Usually phrases forbidding publication without the sponsors permission are included, with the proviso that permission will not be unreasonably withheld. This is much too non-discretionary for most purposes and you should attempt to have some more concrete phrases. Obviously it is courteous to allow the sponsor to see a manuscript prior to publication and CVCP (Committee of Vice-Chancellors and Principals) guidelines suggest a maximum delay of 12 months. Longer periods or indeed indefinite periods should not be accepted. In any event, if a patent priority date has been established, protection should ensue.

The pricing of your contract with a sponsor is often another contentious issue. Is the company making a grant? Is there contract research being undertaken? Is there a joint research project? You must be very clear as to the relationship you are entering. There are a variety of pricing methods that are applicable, but unfortunately universities have used the simplest (and probably research grant oriented) method, i.e. that of cost plus pricing. This seriously undervalues university research and has left us with a legacy that is going to be difficult to shed. Basically, the costs of particular research project are added up and more often than not, that is the price charged. The university often seeks to make its 'profit' (being a non-profit making institution) by swelling the overhead allocation, from between 60 and 120%. Inevitably, some costs remain unaccounted for or not accounted for in full amount. Externally funded research then becomes subsidised by the university, which is clearly unacceptable. Thus if your university's internal policy is to cost contract research in this manner, you should: (a) bear due regard to the true value in the market; and (b) ensure that all costs are recovered. This includes opportunity revenues, i.e. the money you would have gained or lost by doing something else. In the majority of cases such projects will take a considerable portion of your own time, which although you may not like it, belongs to your employer. The rather loose arrangements in universities allow many academics to evade proper accounting of the use of their time and very often excessive commercial activities are not brought into line.

Termination of contract research projects can present two problems: (a) delays in contract signing may cause a lack of money to pay for salaries of staff already recruited; and (b) early termination may also leave a commitment to the university to pay salaries. Thus in any contract research arrangement a suitable clause that covers the continued funding of all non-cancellable commitments (staff contractual obligations) should be included. At the end of a project, there could be some capital equipment that was purchased or donated as a result of the project. It should be clearly stated that the equipment remains the property of the university upon termination. The issues that relate to the charging of VAT on contract research are also complex and you should seek advice. Research as such is VAT exempt, but services are not.

☛ General points concerning collaboration

Informal collaborations can result in unforeseen disagreements and business problems. You should try to engage your partners in a prenuptial agreement whereby you agree upon a course of action for major eventualities. This should include both the worst case failure and the best case success scenarios. In this way you will have an understanding regarding the decision making process, disbursement of costs and liabilities, division of rewards, etc. This agreement should include: (a) the responsibilities of each participant; (b) their financial contribution; (c) ownership of foreground IP and its protection; (d) a strategy for commercial exploitation of new IP; and (e) the legal relationship between all parties and necessary procedures for dissolution. In essence, this contract will be a 'service' agreement with the company and as such you should have it drafted by a lawyer with relevant skills. As R & D workers you may consider that 'collaboration' between like personnel/institutes does not necessarily invoke a special legal relationship. If you do not have any intention of commercially exploiting the results then this is indeed so. However, if you agree a commercial exploitation strategy, execute it and divide the proceeds, then in law you will be considered to be acting in partnership. Similarly, if one party is the sole beneficiary and another is paid to do the work, then there will be an employee/employer relationship. In this case the employee will be entitled to specific protection and the employer will have a special responsibility to the employee. Therefore, you should be careful to consider the nature of any relationship you enter into since it could have important consequences. In any relationship the goals, commitment and hidden agendas can vary considerably between each participant. One can easily see that different individuals, grant giving bodies and investors may all be looking for a different outcome to any given project. If the participating bodies are both incorporated businesses, then legislation relating to competition will generally not apply to R & D projects since these will usually not involve exploitation. It is likely that such projects may be allowed under the European Commission 'block' exemptions, which have been set up to determine whether collaborative relationships are acceptable. For exemption, the agreements cannot restrict the collaborators freedom to operate R & D projects, restrict exploitation by either party, particularly with regard to choice of customers and pricing strategy, and must give all collaborators access to and freedom to exploit the results of the programme. One of the concerns of many in such positions it that of control. Indeed, some are so obsessed by this concept that they prejudice the development of businesses by holding on too tight. A start-up company can be seen as a newly born child that needs to grow up and express itself. In doing so, it will develop a character of its own and eventually assume complete independence from its originator. There are, unfortunately, more disputes over control than anything else. It is often the case that control (which is sometimes synonymous with ownership) will be sacrificed to obtain funding from a third party. In this case, the original owner should take considerable care to choose the new influence wisely. In this regard, a special contractual arrangement should be entered into such that the third party is bound to use 'best endeavours' to commercialise the technology. Again, a specific 'service' agreement should be signed.

☛ Project Management

To the contract should be attached an appendix that relates to the work to be carried out (the schedule). This should contain clear definition of the scope and nature of the work and the milestones that are expected to be achieved. These should be agreed as fair and reasonable by both parties. The milestones should relate to timed events or a programme of work, not to a specific outcome that, because of the nature of research, cannot be guaranteed. Beware of clauses that refer to the cost of failure as part of the agreement; this could mean that the university is obliged to continue support after the termination date. A key point also lies in the frequency of reporting. Ideally there should be a quarterly meeting, preceded by presentation of a progress report in a format that is both acceptable to the sponsoring organisation and which has been discussed beforehand. Always include a half page summary at the beginning, since this is likely to be passed onto heads of department who will not have time to read the whole report.

As with consultancy contracts, there are distinct advantages to putting your contract research projects through the university/institute technology transfer office, compared to arranging it yourself. First, the office should provide experience regarding project costing and IP negotiation. Second, central services such as meeting coordination and secretarial facilities can be provided and third, the office can manage the project in a professional style and ensure the interface with the client is enhanced.

Technology transfer agreements

If you decide that to link up with a technology transfer company is a good approach for you, certain points must be noted. In the initial assessment, the technology transfer company will require access to all relevant technical, patent and business information. You must be prepared to cooperate fully, otherwise the exercise is pointless. However, you should have previously asked the technology transfer managers to sign a confidentiality agreement, preferably limiting access to the information to named individuals only. Your patent agent or solicitor will be able to provide such an agreement, two copies of which must be signed and dated, one to be retained. It is also wise to include some minutes (to be attached as an acknowledged appendix) as to the material discussed.

If and when the technology transfer company decides to take on your case you (and any co-owners, e.g. your institution/employer) will probably be asked to assign the property rights to the company, for which you will receive a share of the revenues. Fixed sums are unwise to negotiate at this point since you do not know the value of your as yet unpatented technology. Since we can assume that commercial concerns are unlikely to pay over the odds for the technology, the other option is that you undervalue it. This could result in bad feeling at a later point. Therefore, it is much better to negotiate a fixed percentage of sales income, the terms of which will vary according to the extent of the services required from the technology transfer organisations, the amount of investment you have already made, etc. For example, if you have already applied for a patent, made a prototype product and tested it then you will have reduced the risk somewhat. Thus you might argue for a higher percentage.

A reasonable agreement to look for would be a single up front payment of gross licence income (which may be up-front fees and/or a percentage of sales royalty) as a means for you to recover your costs (compensation for time, legal fees, etc.), followed by a percentage of the gross licence income whilst the technology transfer company recovers its costs (patents, legal fees, product-development costs). The technology transfer company will incur costs in assessing your technology and you should negotiate strongly for these not to be included in the deductable costs. This is not your responsibility. Would you expect to pay for a test drive if you were buying a new car? Would you expect your department to pay whilst it had a centrifuge on trial? When all costs have been recovered you should argue the share of net licence income that will return to you. Of this money, some may come to you and your co-inventors directly, but this will depend upon the contractual arrangements you have with your employing institution. As a result of the licence, you may be able to justify additional payments for your consultancy service, i.e. those services over and above what is necessary for transfer of the technology. Everyone knows that no matter how accurately and precisely a technique is written down, the help of the originator or one familiar in the technique is of great benefit in smoothing the passage of the technology through this stage. As a result of the licence you may well have some stated obligations and your time will be part of the negotiations with the licensee. Indeed, you may be specifically named.

Once the agreement with the transfer company is signed, you will probably join the negotiating team. Your presence as a scientific font of wisdom and your knowledge of the field, together with ideas for new applications and markets will be invaluable.

Industrial sponsorship

Having originated an idea and perhaps performed sufficient research to make you believe it has commercial potential, you may consider patenting. As we have discussed, this process is long and expensive and your idea may not have sufficiently wide applicability or novelty to warrant this. This does not mean that it is of no value. Transfer of your technology to a commercial organisation could reap you some reward in terms of consultancy or valuable contacts for future projects. Major shifts in UK government policy are now directing grant holders towards more interaction with industry, and an understanding of the workings of commercial research and development is crucial. However, many projects may require development funding before commercial potential is realised and others may require a more concerted research effort to even begin a development project. In addition, commercial organisations often have access to capital equipment that maybe otherwise unavailable to the researcher.

The research culture in industry is becoming increasingly more competent and the intellectual 'industry versus academia' divide is closing rapidly. Thus joint research projects between commercial concerns and academic groups, such as universities and charity funded laboratories, are becoming common place.

Allocation of company funds

If you are dealing with a large public limited company, funds for research and development projects will come out of the internal budget for the department concerned. Some companies have a department for external affairs with their own budgets for participation in high profile government schemes (e.g. LINK) and some will give to charitable institutions, which in the healthcare section will fund their own-named research laboratories. The latter is almost inevitably resourced out of pre-tax profits and is thus a **board** level decision, since these are monies that could otherwise pass to the shareholders. Smaller privately owned companies allocate funds at the discretion of the owners, probably via a recommendation from the managing director. These companies tend to be 'technology hungry' and are cash limited, except perhaps immediately after financing or refinancing. In general, in such companies resources must be efficiently channelled towards internal projects and the scope for external liaisons is more limited.

The enterprise initiative

The enterprise initiative is an all embracing package of help and advice provided by the Department of Trade and Industry (DTI). This service is available to all business sectors and, with an open mind, you will find that much of the advice is relevant to the biotechnology sector. This includes: (a) a consultancy service (if you have less than 500 employees); (b) help in improving/creating competitiveness; (c) advice on the single European market, exporting, environmental issues; (d) assistance in forging links with educational establishments (i.e. someone may approach you!); and (e) help for businesses in regions where there is an ongoing urban development programme. The DTI runs an innovation enquiry telephone line whereby you can obtain literature on DTI schemes, but this route is not appropriate for technical advice. For the latter, there are a number of regional offices that may help.

The first approach is to telephone your local DTI office, explain your needs and obtain relevant literature. This could be followed by an initial business review with an experienced DTI appointed person who will decide whether you need specialist help. If you are proposing a biotechnology start up or have a business development problem in this area, you will undoubtedly need such a specialist. The initial meeting is free, and the enterprise councillor will send a report to a specially chosen contractor who will find an appropriate consultant. In this second stage, the government will contribute towards the costs of the consultancy to 50% for between 5 and 15 days work. The contractor will review the programme of work and help negotiate fees, so that you have a reasonable chance of getting value for money.

In particular, this scheme can help you with business planning (review of objectives and strategy, changing markets, etc.), the design of your product and manufacturing strategy, installation of management and financial information systems (databases on customers, competitors, presentation of accounts, etc.) and the complex issues associated with producing a quality product. You will need rigorous quality control of production, product data sheets, etc. The final area where consultancy help is necessary is in the marketing of your product or service. It is absolutely essential that you understand the needs of your customers, since without satisfied customers sales will not accumulate and profitability will be unattainable. In addition, you must research and understand the competing products, making sure you know where your competitors are going. The scheme will also give you access to advice on the pricing and distribution of your products and how to arrange an after sales service. As an example of the latter, a reputable company recently sent out a vial containing 800 μl of antibody, when the catalogue, order and invoice all clearly stated 2 ml. They were unaware of this until it was pointed it out to them. What was their response? By courier, a new vial of antibody arrived, this time with the correct volume and a note of apology (NB: without admitting liability). Would your response have been similar?

As a note of caution, the consultants appointed by the DTI are all registered, but may be found to be variable in quality. Make sure you check their credentials by more than one independent source before agreeing to use them. If they are of the required standard, they will not only be expecting this approach but will encourage you to take such references and will provide additional names. Be cautious of those who show reticence in this direction.

In order to provide support to small and medium size businesses, the government has instituted a number of regional 'one stop' shops where you can go to obtain business advice. It is a partnership between the DTI and business support organisations, designed to give success to the wide range expertise that you are likely to need. Fees will be payable, so this may be more appropriate after the start-up stage.

For small businesses, there are two areas that are worthy of mention. Ideally, you should have already formed your company by this point.

☛ Grants for R & D

These are given to aid the introduction of a project into the market, primarily by: (a) reducing the time to completion of a project; (b) reducing the risk in a product development programme, either technically or commercially; (c) enabling a programme to be initiated and run, when otherwise this would not have been possible; and (d) enabling more effective use of your existing resources. These grants are an excellent way of helping to fund your business in the early stages and can give you the means necessary to get the business off the ground. Receipt of such awards is also prestigious and your credibility will be increased if these have been obtained.

The SMART award is a nationwide annual competition for the development of commercially viable products, from the research stage. This is for a maximum of £45 000 as at 1994, an additional third of which must be provided by the company. This means that the likely budgets will be around £60 000, with you having to find £15 000 from other sources, which could be internal cash generation, external investment from a seed venture fund, your own money and so on. SMART winners will receive up to a third of the award upon the initiation of the programmes, i.e. a £45 000 grant means you will receive £15 000 on day one. This is a considerable bonus to a company that may have very little spare cash, and this injection of finance at a critical point can be the difference between survival and disaster. The applications are usually required by early April and the programme is announced some two months before the closing date, by your local DTI office, from whom the details can be obtained. If the results of your programme look promising, then in the final quarter of the SMART award you may apply for a second SMART award (SMART 2) for a larger sum of money and the progression of your project nearer to the market place. The DTI accept that there is a significant degree of risk in your project, being essentially of a research nature, but the scheme should not be viewed as a way to disguise an academic research proposal. First, the project must have the marketing of a product as a primary objective and preferably one that will enhance UK competitiveness. Scenarios in which the technology is intended for licensing are less likely to succeed, i.e. you should produce and market the product from your own company. Second, in your proposal you will be expected to realistically indicate the risks involved and estimate the probability of failure. This should be given good attention, as this is not an easy section to write. One further tip is to pay particular regard to your project management scheme, balancing the design, development and testing phases, and making sure the programme is within the capabilities of the budget. A £60000 project is not likely to involve more than two people full time, so the project should reasonably be completed in not more than two designated man-years. The guidance notes obtained from the DTI are excellent and should be both read and followed carefully, in order to maximise your chances of an award.

Two other schemes are available, know as SPUR and the Regional Innovation Grant. SPUR is again a scheme for small businesses, i.e. those with less than 500 people, to develop new products and processes for the industry sector in question. These tend to be applicable to more mature businesses, rather than the new biotechnology start ups, but could be considered after two or so years of successful trading. The award is for 30% of the eligible costs up to a maximum of £150000. Clearly, the remaining 70% must be financed from other sources, so you need to be in a financially well-developed environment. The Regional Innovation Grants are available for small businesses (> 50 people) to develop, improve and introduce new products into the market. The grant is for 50% of eligible costs up to a maximum of £25000. It is thus for you to raise a matching sum from internal resources, whether this be in cash or an external contribution from, e.g. an industrial partner (a joint development project) or venture capital fund. The grant is payable when the expenditure has been made, so you must have sufficient funds to finance the project in the interim. However, if the project exceeds three months in duration, an intermediate claim can be made on a pro rata basis when at least 35% of the costs have been expended. There are certain geographical restrictions on the award of a Regional Enterprise Grant and a map of designated development and other assisted areas can be obtained from the DTI. The application procedure is relatively straightforward and, as for a SMART entry, requires a project proposal with an assessment of the degree of technical risk, a business plan with realistic cash flow projections and the latest account, if available. More than one innovation grant may be obtained providing this involves different projects but it is unusual to receive a grant for more than one project at any given time. If you have already received a grant, it will be highly advantageous to have commerically exploited the results of this grant, before applying for another.

☛ Collaborative research schemes

Providing you can make contact with an industrial partner, there are several schemes by which commercially orientated research and development projects can be considered. These are an excellent method for gaining momentum for your business, not least of which is the opportunity to forge lasting industrial links. In particular, there are predefined advanced technology programmes in which you may participate and the best approach is to speak directly to your local DTI office in order to discover exactly which programmes are relevant to you. The remit and expertise of the DTI is very wide and you can expect help in finding suitable research partners. A grant for up to half of the eligible costs is available and there must be at least two commercial partners, one of which must be in a position to exploit the technology. Two other schemes require mention, those of LINK and EUREKA. Within the LINK directorate there are a wide range of technologies, several of which are in biotechnology. The DTI and research councils provide half of the total project cost, the remainder being divided between the industrial partners. The scheme may have one or several partners, but as the numbers increase so do the complexities of managing the intellectual property position. In essence, your university department can enter into a LINK arrangement with industrial partners, perhaps with the expectation that the academic partner will retain some rights to commercially exploit the technology at some stage.

The European dimension is also available, both with DTI (EUREKA) and EU funding. The former has many active projects involving a plethora of technologies and an active support system to engender contacts, idea generation and networking. The project accords status and will receive considerable publicity, which could be of value in the development of future contacts. However, if using such as scheme one must be careful to maintain commercial confidentiality. There is not a budget for EUREKA funding *per se*, but since responsibility for the participation of UK companies rests with the DTI, some degree of financial assistance may be available. This might include a contribution towards assessing the feasibility of projects, finding and arranging projects and, if you are a company with less than 250 employees and modest turnover, there may be assistance with the project costs. Criteria for eligibility are wide and require: (a) demonstrable managerial and technical capability; (b) a project that could lead to a significant technical advance in a high technology field, specifically for a new service, product or process; and (c) at least two participants from different EU countries. To apply, the first step is to send a short summary to the EUREKA office to include the nature of the product, and the extent of the technological advance, the size of the international market for the product and when you think you might get it there, the costs to you and your potential partners and the identity of the partners. If you are still seeking the latter, there is a EUREKA networking facility, which contains a partner search opportunity and indeed, if you would like to examine the nature of the projects underway, there is a public database available. Further details can be obtained from the EUREKA enquiry point at the DTI.

The DTI also provides help in the technology transfer arena, specifically for a life sciences company in the 'Advanced Sensors' (bio-sensors) and 'Biotechnology Means Business' programme and, in both cases, information can be obtained regarding commercial opportunities in biotechnology. In addition, there are 12 Regional Technology Centres that will help in technology transfer services.

These facilities should be highly advantageous to the small biotechnology start up, first by forging links with the local, national, European and even international business community for you to sell your services/products, and second by giving you access to technologies that may complement your own portfolio.

Considerable help for business development is available from the European Commission (EC) under the Fourth Framework Programme (1994–1998) for research and technological development (RTD). The EC offices provide a guide to research funding, a regular newsletter (RTD–INFO) and access to a series of Comunity wide RTD liason offices. Information is also available from a dedicated help desk and an on-line computer database (CORDIS). Via CORDIS, you may also obtain information on the Value scheme, which enables small- to medium-sized enterprises (SMEs) to apply for funding with regard to fact-finding missions and technology transfer. SMEs are defined as having less than 250 employees, annual sales of less than £15m or assets of less than £7.5m. These systems are backed up by giving access to venture capital funds, particulary seed capital and eurotech capital, which will be discussed in Chapter 8.

The CRAFT scheme

The European Union has brought forth a scheme that will be of considerable benefit to smaller companies. This scheme, CRAFT, is being emphasised in importance by the DTI and a large body of funds is available for research and development projects undertaken by small- to medium-size enterprises (SMEs; which are generally defined as having less than 250 employees, annual sales of less than £15m or assets of less than £7.5m and less than 25% share ownership by a larger company). The network of SMEs makes a significant contribution to the exploitation of new technologies. The EU has renewed the CRAFT scheme in The Fourth Framework Programme such that SMEs can provide financial assistance for research contracts with universities and research organisations. At least four SMEs from two EU member states are required each of which must contribute 50% of funding, the rest being provided from EU sources. The recipient of the grant is in a fortunate position since up to £500000 can be obtained, with 100% funding. There are even mechanisms to finance proposal preparation. The majority of projects are in lower technology sectors when compared to biotechnology, but there is no reason why applications should not be put forward from the latter sector. There are two potential ways of taking advantage of this scheme; first as an academic laboratory that runs a research project for SMEs, which will give you contacts at the very least and second by using your biotechnology start-up company to participate in the scheme. The obvious problem is in partnering, i.e. participation in the project. Given the cash limitations of SMEs, their tight focus and specific needs, a considerable amount of effort will be required in order to identify suitable partners. Fortunately the

DTI are able to offer help via a partnering service and you should contact your local Business LINK office for details.

Exporting your products and services

At some point in the development of your business you will need to consider exporting the services and products that pertain to your company. The global nature of biotechnology means that an understanding of this process and its effective implementation will be essential to your commercial success. These are issues that should be considered at an early stage and built into your conceptual framework of the business, long before you write the business plan. The export strategy will then become an integral part of the business plan and indeed, investors will expect to see a formulation of your intentions. The DTI produce three booklets (these can be obtained form your local DTI office) which will be of assistance and that document the services and help available. In order to use these services, you must be a UK based company intending to export.

In the first instance you may be able to seek advice from an export development advisor or export promoter, and details of these persons can be obtained from your regional DTI office. However, this step may not be necessary as you will probably already have a relatively firm idea of the markets that you wish to enter. Clearly, at this point you need to have quantifiable data regarding your prospects of selling in a particular geographical, market and the best way of doing this is with a market information enquiry service that will provide you with a tailored report on any export market in a short period of time, to include assessment of market prospects and the relevance of your product or service to them, general political, legislative, tariff and economic information, answers to specific questions and recommendations on how to proceed. Considering the amount of information you are likely to receive, the charges are relatively modest. The next scheme of interest is an export marketing research scheme (EMRS), which is managed by the local chamber of commerce. This is a highly subsidised service to support acquisition of market data and again, you are advised to contact the DTI since the rules for grant of subsidy are precise.

If you are searching for an overseas agent or distributor then you will need the export representative service (ERS). The use of representatives is surrounded by a complex legal web, largely to prevent exploitation. Dismissing an unsuitable agent can involve court action, and the DTI will help you through the potential pitfalls so that you are less likley to appoint an unsuitable agent in the first place. Application for the service involves the completion of a fairly long and involved form. This should be given considerable attention since its purpose is to help the DTI post in the designated country find you the best possible agent. The more information that you provide, the more definite you are about the product/ services you wish to offer and the more certain you are of the type of agent you require, the easier it will be for the overseas post to find you the agent that you need. The search for an agent takes between 8 and 12 weeks, with a rebate being available on an already highly subsidised service if you follow up the market information within six months of receiving the DTI report.

There are a number of other services which may be relevant:

1 The overseas status report service. If you are introduced to or approached by a foreign company you need to investigate the credibility of this company. This service can help you decide if the business/distributor/agent is right for your purposes.
2 New Products from Britain Service. If you have a product, process or service of UK origin that is of newsworthy calibre, then this DTI service will help you promote it overseas by giving editors of relevant magazines access to relevant information. Acceptance of the story is at the discretion of the particular editor, but you have the advantage of your story being presented along appropriate channels and written by professional journalists who are experienced at targetting particular publications.

Why export at all?

In the case of biotechnology associated products we must begin with the assumption that we are entering the European arena at the very least. However, many are now in the position of having to explore the much larger EU markets in order to boost sales. For most products and services, the profit margins that can be obtained by selling in the EU are either equal to or are in excess of those that can be obtained in the UK. Unit costs will be also lowered (hence also contributing to higher profit margins) since stocks can be effectively reduced. Having your firm trading overseas will also enhance your company image, give access to new ideas on quality, marketing strategies and product presentation, which can help in the home market. Furthermore, overseas operations help to spread market risk. If one market is suddenly depressed due to changing circumstance (e.g. an unfavourable patent ruling) then this can be buffered by the other markets in which you trade.

If you intend to enter a non-UK market then your marketing plan (which is now likely to be an addendum to the business plan) should contain a specific export plan. This should consider the necessity for an additional budget, the time of payback, how much management time will be required, how your product is likely to fare in the chosen market (quality considerations, position relative to competitors) and the requirement for staff training (languages, documentation, etc.). It will be necessary to specify the currency of payment, take safeguards against currency movements and to consider alternative insurances against non payment. A fundamental decision will be whether you wish to sell to the new market directly or appoint an agent to do this for you. Both routes have their pros and cons, which will have to be considered carefully in the export plan. Direct selling will require a lot of management time and indeed, a skilled and experienced manager. An agent can represent you in the market place or may hold and distribute stock on your account. If you end up with a poor agent, then this effectively closes a market to you. Additionally, there are considerable legal implications to agency arrangements of which you should be aware and you are strongly advised to seek advice from the DTI export initiative before proceeding. When considering an agency, make sure that you have consolidated your trademark position in the territory of the agency. If you have not done so, then it would be possible for the agent to register your trademark and use it without your consent. It can be an expensive and lengthy legal procedure to recover such a mark.

As with any other market exploration, you should first perform some market research. To facilitate this, there is a national Export Market Information Centre. This resource contains market research reports, overseas directories, development strategies for particular countries, the ability to search commercial databases and world trade statistics. It is available as a self-help information centre. The DTI will also be able to provide help with individual national legislation regarding duties, local taxes and exchange controls. Similarly each country has technical standards which must be met and again, details can be obtained from the 'Technical Help to Exporters' Service via the DTI.

If you intend to trade outside the EU, then there are two major markets that you should consider with vigour. The first is the sale of your products and services in the US, which is, rather obviously, the largest English speaking market. However attitudes and working practices are considerably different on the other side of the Atlantic and you should try to find a contact who can guide you through the US system. In particular, the legislation for exporting particular products to the US can be complex. Devise a US marketing plan and consider the use of an agent in the first instance. The DTI export service in the US is particularly good and will provide a useful starting point. It is advisable to wholeheartedly use this service since the complexing of agency legislation in the various US states and the prevalence of court actions will necessitate that you acquire guidance and/or protection. Given the high number of biotechnology companies in the US, this market is a good arena for entering into a co-marketing arrangement with a like minded company. There are several US based bio-partnering conferences per year and it is wise to allocate some of your budget to sending one or more of your staff to such an event. Present at such meetings will be representatives of the venture capital community, business angels and business representatives looking for suppliers, customers and partners. If you can find a company to represent you in the US, then you should jointly devise a marketing strategy which will become a schedule in a full agreement. Many US companies are looking for a European distributor and the UK is an obvious stepping stone for many of them. Thus the path of co-operation rather than competition maybe advantageous.

If you wish to enter the Japanese markets, then it is almost essential that you develop a relationship with a Japanese 'trading house'. There are some eight major Japanese corporations, many of which have a wide portfolio, who will undertake a technology transfer/commercial relationship on your behalf. The structure of Japanese business and society is such that you will only be able to approach Japanese customers via one of these trading houses. There are now several texts and seminar series which will enable you to learn about doing business in Japan.

University technology transfer and industrial liason in academic institutions

Many universities/institutes are now developing their offices of technology transfer. Some universities, notably Oxford, Imperial College and University College, London have an excellent record in this area, but many of the UK's (and indeed Europe's) high class educational establishments are severely lacking in adequate commercialisation skills and to date have resolutely failed to properly exploit much of the research that has been undertaken. There are three major problems. (1) The offices of technology transfer are severely under resourced. In some cases there is not even a fund for initial patent applications, let alone project development, licensing or business start up. (2) The staff employed are rarely of sufficient stature to carry the job. Inexperienced in business matters or highly experienced and at the end of their careers, the staff often do not have either the skills or motivation necessary for the correct fulfilment of their roles and the high expectations placed upon them. (3) There is a credibility problem. If academics do know about the office of technology transfer, they tend to look down upon the efforts of such staff. This is usually because the staff employed are not of sufficient academic stature for productive interaction. Hence the sources of new technology dry up since there is a tendency to think 'I can do better myself', which is, alas, sometimes true. Also, these same staff do not generally have the business stature to deal with the decision makers from industry. Whilst it is true that there is a shortage of suitably qualified individuals, it is also true that insufficient attention is given to selection of the candidate and to structuring the teams necessary for efficient transfer of technology. The functions necessary are (minimally): (a) two people who are involved in the search and appraisal of new technologies in (i) the life sciences and (ii) the non-life sciences, engineering, chemistry etc.; (b) a patent/licensing manager; (c) a business start-up manager (financing and organisational issues that pertain to a business development); and (d) a marketing manager. Together with administrative staff and a business development manager to oversee the whole process, around eight people will form an effective team for a medium to large UK university. Clearly, a salary bill of around £180 000 plus costs (patenting, marketing expense and overheads) is outside of most university funding aspirations, where valuable resources are more often required for teaching commitments and research. However, the returns to a university for sufficient resourcing of a proper technology transfer effort could be significant, depending upon the research output. Those universities that are destined to become one of the chosen few 'research' universities will of necessity have such an office as a central component of their competitiveness.

The idea that technology transfer offices are 'commercial' and must therefore be able to 'afford it' is truly fallacious, and university authorities should begin to look at their technology transfer office as an investment, not a cost. As with any business, it may run cash negative for a period, but the risk and reward mentality must apply and as government policy rationalises grant income, universities must wake up to this important source of revenue that has hitherto gone to waste.

At least 36 UK universities have a technology transfer capacity, which are formed into the Universities Companies Association (UNICO). This organisation is very young, but it does provide a platform for the dissemination of ideas and experiences in the technology transfer arena. Similarly, there is an organisation known as the University Directors of Industrial Liaison (UDIL) whose aim is to promote professional relationships between industry and the universities. In particular, the representative of your university should be able to advise upon research funding, commercial exploitation, contract negotiation and marketing. The CVCP recommend that each institution should have a central office for the management of technology transfer and most importantly, for ensuring both internal consistency and broad compliance with agreements involving other institutions.

There has recently been a technology audit of the research in UK universities, carried out by the DTI. Few universities appeared to have the critical mass of technology that is necessary to guarantee the success of a technology transfer office, perceived as a high risk venture that will not provide a return for a long period. In addition, many universities did not consider that technology transfer represented a core part of their business (research and teaching) and furthermore, they did not have sufficient internal resources to exploit and promote the existing technology. This shows a lack of commitment to both the reforms that are facing UK universities and the technology transfer ideals. This has to change if a significant position in the high technology sectors is to be both assumed and maintained; at the moment we have an incredible loss of national resource, as trained staff move overseas for better opportunities and many good ideas are unexploited, only to be discovered and patented elsewhere. Indeed, even a recent reference to a 'technology audit' was considered to be too threatening and 'opportunities review' has been suggested instead!

In any laboratory or indeed commercial settings, there are many stories of lost opportunities and potential products that have failed to be developed, e.g. lack of motivation for the refinements that are necessary for a commercial product, failure to recognise the commercial potential, a lack of a company that is willing to invest the relevant finance and absence of necessary complementary products.

The technology transfer office cannot solve all of these problems but it does have the potential to solve some of them, perhaps by finding a commmercial sponsor who is willing to invest the relevant capital or by assisting in the company start-up process. Between each university and institute the calculations applied to income and expenditure vary somewhat, but in general these offices do not make significant monies from licensing income especially when the figure is compared to research grant income. The 'big hit' licensing agreements are rare and indeed, within the university sector licensing arrangements do little more than cover the costs of execution. However, licensing income is not an appropriate measure of the effectiveness of the technology transfer function and a more applicable measure is perhaps the university income that is lost if a technology transfer office is not in place.

The technology transfer process

The technology transfer function should consist of:

1. An interview with the inventors to assess the technology and make an initial decision as to protection of any intellectual property.
2. The production of a non-confidential summary.
3. Identification of potential clients/customers via database searching and networking.
4. Decision as to licence the technology or form a start-up company.
5. If licensing is chosen, perform market research to determine the level of interest in the targetted companies.
6. Execute confidentiality agreements.
7. Supply the confidential information.
8. Follow up the contacts to ascertain interest and obtain a definitive commitment, ideally to Heads of Agreement.
9. Negotiate a licence and monitor the transfer of the technology, including collection of royalties.
10. If a start up is chosen, prepare a business plan, identify sources of finance, implement programmes for marketing and a product development.

The remaining chapters in this book will introduce at least some of the issues that are associated with each of these functions, beginning with licensing.

6 The licensing option

But what is freedom? Rightly understood, a universal licence to be good.
Hartley Coleridge
In *Liberty* 1833

Introduction

A licence confers permission to do something which would otherwise be constrained. There are many forms of licence such as a fishing or driving licence and all have their own legal characteristics. However, licensing should be considered to be the tool of commercial practice and the route to mutual profit. To an organisation, licensing is a part of the overall activities of the company and cannot be separated from it. As such, licensing represents one of a series of options to a potential partner and this should be borne in mind once you have decided to licence your technology. Licences form an alternative to in-house R & D programmes and can be used to completely buy in a new technology or to augment existing programmes. Theoretically, the latter should maximise the use of internal resources without total dependence upon outsiders. Additionally, licensing can be used where direct capital investment in a project is difficult for political or other reasons. The action of licensing implies a degree of trust between the parties and could well be a long-term relationship, with the generation of a considerable amount of goodwill. Relationships can break down and you should take this into account when entering a licensing arrangement. Above all, when negotiating a licensing arrangement, proceed *independent* of trust.

Why should you grant a licence?

The major reason for licensing your technology will be to generate revenue for yourself and/or your institute. This will be in response to perceived market demand for a product or to enable a licensee fill a gap in a research portfolio. If the licence is to be handled by your technology transfer office, then there may also be good political and resource reasons for entering into a licence agreement. In order to maximise revenue it is likely that by granting a licence you will be able to take advantage of the marketing, sales and general support infrastructure of a larger company. You would probably not be able to provide these resources in the short term, if at all. Under some circumstances, you could be compelled to enter a fair licensing agreement if it was considered by a court that you were making unfair use of a monopoly position.

It is important that the inventor is confident that the arrangements for licensing are of a fair and reasonable nature. In general they will wish to remain involved with the process and will be invaluable to the technology transfer process at a later point. Hence, both during negotiations of the licence and after signing, it is important that there is a good working relationship between the inventor, the university/ research institute (usually the licensor) and the licensee.

Why should a licence be purchased?

Looking from the point of view of a potential client, the holder of innovative technology should consider the reasons that underlie the decision to purchase your technology. Try to find these out during negotiations, since they can give you some additional leverage. R & D teams are very expensive to both set up and maintain, and for a large organisation their contribution to the overall effort is difficult to quantitate. Licensing technology from universities can supplement internal company resources, usually at a much cheaper rate than could be obtained from a commercial licensor. The action of licensing in the technology thus ensures the leading edge nature of the R & D department and if this is done both diligently and selectively, it decreases the risk on future product development. Existing resources can then be channelled towards the developmental aspects of the product or planning for the next generation of products. Another advantage is that of using technology that has been proven already, whereby many developmental obstacles should have been removed. The latter can result in considerable savings in project completion times. If these benefits can be extended towards taking the product to the market, then there is an additional reason for entering a licence agreement. It is difficult to quantitate such advantage at the 'early research' stage where the risk of failure is high and where the market is untested.

What are the types of project which require a licence?

One can envisage several types of project that will require a contractual arrangement between yourself and/or your institute/university and potential sponsors or licensees.

1 A straight transfer of technology. In this case you will be attempting to licence the right to use some technology or reagents, with or without obligations on yourself

to provide additional services. For this you will be attempting to secure an upfront payment plus a royalty. A licence agreement will be applicable.

2 A joint venture. Here some additional R & D may be necessary. Whilst a joint venture between industry and a university is likely to make the former uncomfortable, there could well be a joint venture between an industrial partner and a small start-up company, perhaps one in which you have a significant stake. In this case there could be a requirement to use intellectual property rights and again, a licence agreement will be necessary.
3 Contract research. Here an industrial concern may require a piece of research on a particular subject. They will be funding a group in the university/institute or business start up and will require the results. During the course of this research, foreground intellectual property could be generated and presumably access to some of the background intellectual property will also be required. Hence an appropriate licence agreement will be necessary here also.
4 Consultancy and studentships. These should not require a full licence agreement but nevertheless require an agreement, to protect intellectual property issues and ensure appropriate payments.

Initial negotiations and the non-disclosure agreement

Let us now suppose that you have identified a series of potential licensees. The use of the plural is deliberate and you should always try to keep your options open until an acceptable agreement with any given partner can be reached (i.e. a signed and dated binding licence agreement). The source of these licensees could be from personal or colleague derived contacts, searching any of several biotechnology indexes, responses to your entry in relevant databases or a proper market research survey.

The first approach should be a friendly (and brief) telephone call to ascertain interest; ensure that you do not give away confidential information. If necessary, write down what you are prepared to say and do not exceed this. Try to make sure you obtain a name, preferably that of a head of department, licensing or contracts manager and speak directly to this person. If this person protests their irrelevance, try to get a name and telephone number in a relevant department. If you have a positive response, the next step is to send a non-confidential summary with a covering letter. The summary should be brief, no more than one A4 side, written in semi-technical terms and should immediately convey the invention, without revealing it to one 'skilled in the art'! Ideally, you should check the document with you patent agent so that you do not inadvertantly reveal confidential information. The covering letter should also be clear, brief and concise, stating who you are, that the technology is available for licensing and that in order to proceed a non-disclosure agreement should be entered into. Do not allude to contract terms or money at this stage. Your summary will hopefully be bounced around a committee (or two) and/or end up on the desk of a decision maker who will ask for further details. These should be sent only upon receipt of a non-disclosure agreement (NDA), i.e. send two copies signed by someone with relevant managerial capacity and wait for one to be signed and returned. You will find that many major companies will not be compromised by receiving confidential information, i.e. they would not wish to be accused of using information received, when perhaps they were even more advanced in a similar research programme! Thus do not send unsolicited confidential information and save everyone embarrassment.

Upon receipt of the NDA you can send the relevant confidential information, whether this be a copy of your patent application or whatever. It is imperative that you send this by a recognised courier in a timely fashion. This will give a guarantee of receipt and a proper impression of your professional approach to technology transfer. Also, specify a date for a reply and for return of the document, using the same means of delivery. The review process could take a considerable period of time, especially since some companies will send patent applications to external agents. If you have not heard after two months (make a note in your diary), then make a polite telephone enquiry. If you get a positive response at any point, the next step will be to arrange your first negotiation meeting.

Negotiating the licence

We have already discussed the tactics and styles that you may adopt in negotiating for the commercialisation of your technology; it is now appropriate to consider the negotiation. The errors are most likely to come in what you leave out rather than in what you remember to discuss, so hopefully you will be able to tailor a checklist from the following discussion.

If you intend to use the university/institute office of technology transfer, approach them at the earliest stage and keep them informed. There is little to be gained by withholding information even if you consider it to be irrelevant. The most important practical pieces of advice are to: (a) have patience; (b) be silent when necessary (and sometimes when not); and (c) be prepared to walk away from the deal if it is not what you want. A 'sell at all costs' attitude is likely to lead to a disastrous deal and if your technology is at all interesting, you will find another potential licensee quite soon. Ideally you should play one against the other until you get a sensible licensing position.

The negotiations are likely to mostly concern business issues, with perhaps discussion of a few scientific and technical points as necessary. Be extremely clear as to how much information you are going to reveal, i.e. make sure your negotiating team has predetermined the extent of this information. Similarly, discuss a few business scenarios beforehand, so that you can contribute intelligently. Commercial companies do not wish to hear academic presentations unless this is specifically requested, since the latter often become requests for research grants. Whilst under some circumstances this may be desirable, it should not necessarily be the outcome of a licensing negotiation.

Pre-licensing negotiations and confidentiality

The further negotiations proceed, the more information will of necessity be disclosed. Academics are a notoriously easy target for revelations of their research, being captivated by an all encompassing enthusiasm and desire for recognition. The utmost caution is recommended, including a significant amount of pre-negotiation planning with a legal advisor or business development manager. In this way you can gain a perspective on exactly how far you can go without compromising what could be a valuable licence. A word or two out of place in response to a goading question could give away a key piece to a puzzle, so restraint is the order of the day. The first stage will involve the initial approach, expression of mutual interest between the parties. This could be from an interaction at a conference, from being an invited speaker or from old colleagues. At the first meeting, one should establish the ground of mutual interaction and generally introduce the respective institutions. Both parties should then investigate the practice of the other with regard to entering licence agreements and general licensing practice. In general, the more agreements that have been entered into, the more likely that there will be a pool of relevant experience and a rapid conclusion to the process. In addition, some business information such as potential sales of licenced products may be of consequence and may be revealed during such discussions.

Thus, these initial stages are usually a non-confidential exchange of information designed to determine if a licence between the two parties is appropriate. At the next stage more information will be transferred and confidentiality agreements will be required. This will still be of a general nature regarding market information, (e.g. costs, competitors), legal information, etc. To avoid embarrassment it would still be wise not to receive or disclose sensitive commercially relevant information. The confidentiality agreement should only encompass enough information for operation of the licence; other information will fall outside its scope and will presumably not be part of discussions. The fundamental tenet of the agreement will be an obligation not to disclose this information to third parties and preferably it should be: (a) confined only to those employees who need to have it; and (b) acknowledged as being received by the individuals concerned. The licensor should insist on knowing the identity of these people. Since not all the information being transferred will be confidential, that which is must be identifiable and specified as such. There will of course be exemptions to this, notably: (a) information that is already in the public domain; (b) information that subsequently comes into the public domain post signing the agreement (except by default by the licensee); and (c) information that may already be in the licensees possession. It is essential that you establish exactly what position they are in at the outset and, if possible, determine their research direction. Even if the licensee is unwilling to give this information directly, it can usually be obtained one way or another.

Obviously one cannot restrict the use of information that the licensee already possesses, but you can restrict the use of the information transferred, i.e. you should be careful to state that it is used for a specific purpose and no other.

Upon termination of the licence, you may be able to stipulate return of the information (e.g. a plasmid) but in most cases this will be impractical. Post termination, certain licence clauses will survive the agreement, breach of which could result in legal action. Exactly what these are will depend upon your individual circumstance and should be subject to negotiation.

It is most important to address the role of third parties in the confidentiality arrangement, these being for example, sub-contractors or employees of the licensee. It is likely that reasonable non-competition clauses will be acceptable to the judiciary, but in general they will not approve of any restriction on employees activities.

The informal letter

In the first instance you are likely to reach an informal oral agreement regarding the transfer of IP. This is of course very difficult to prove. In the second instance you may exchange letters which outline the discussion that you have just had. These may be in a form that is to be countersigned and retained and you should be cautious of the wording of such a letter, since if accepted by the licensee it could constitute an agreement. Obviously you can add clauses that preclude such a conclusion by making it dependent upon a formal licence agreement, etc.

Heads of Agreement

Following negotiation, one or other party will draw up a preliminary document that contains the general principles and direction in which the arrangement is likely to go. This is known as the 'Heads of Agreement' and is an extremely important document that sets the tone of your future relationship. One or other party will take the responsibility for drawing up this document. Ideally, this should be you, in conjunction with your legal advisor. It is imperative that you maintain the momentum (leaving this to the licensee could result in delays) and set the tone of agreement to what you want or have agreed. It is not unheard of that one of the parties writes a Heads of Agreement that is dramatically opposed to what you have previously agreed verbally. The temptation of course is 'to let someone else' deal with it for whatever reason, be these other priorities, a lack of familiarity or prohibitive costs. It would be wise to reconsider the consequences if you decide upon the latter route. Politically of course, if you are a university/research institute based researcher, dealing with a biotechnology or pharmaceutical company, it is easy to get swept away by the smooth professionalism and the knowledge that they generally have both enough money to cover the costs and a number of experienced legal personnel. Expenditure on such commercial matters may be hard to justify within your employing institution, which may not have a budget for such matters and may be less experienced in commercial dealings. The prospect of first having to fight an internal battle for funding can indeed be prohibitive. If the finances to cover your legal costs are not available for drafting, then of course you probably have little option but to save your funds to use on having a qualified person read the final agreement. Larger companies, such as those of the pharmaceutical industry, will often insist on drafting the agreement and whilst you still have to be careful about conceding too much, especially relating to future discoveries, you can be sure that the agreement is likely to be legally correct. In general, such companies deal with so many technology licensing agreements, that you are likely to get a tried and tested agreement. This is not to imply that it will necessarily be acceptable to your needs, but may provide a useful framework with which to negotiate your own particular circumstance and technology. As for the informal letter, the Heads of Agreement should be kept brief, to the point and unambiguous, without any specification of how final the document is. The Heads of Agreement should contain: (a) notification of the parties involved; (b) the nature of the technology under consideration; (c) the scope of the licence regarding geography and exclusivity; (d) the fees (upfront payments and royalties) to be paid, how and when; and (e) any other relevant commercial points including major guarantees or warranties, third party considerations and general procedures for arbitration and termination. Again, add a clause stating that the Heads of Agreement are for discussion and do not constitute a formal offer.

The potential licensee

Before negotiation, undertake some **due diligence** on your potential client. Obtain company reports and accounts, together with catalogues (useful for product profiles) and any other information you can muster. Analyse the financial status thoroughly; be particularly wary of immature, technology hungry biotechnology companies. Often venture capital funded, these businesses are loss making in their first few years and many never pull themselves out of the trough. Whilst you can cover for eventualities such as bankruptcy or inability to take the product to market, you could loose valuable commercial time and the agreement could prevent you from routing your technology through someone who could commercialise it effectively. In order to ensure action, you must quantify the 'best efforts' clauses, e.g. guaranteed minimum royalties after a fixed time period. If your client believes their own sales predictions for your technology this should not be a problem. Your contact in the client organisation must also be vetted carefully. Find out their status and whether are not they are: (a) a decision maker; and (b) have authority to complete a deal.

Project management

Some licence agreements will not just be a transfer of materials and technology or a licence to use, but also a joint research and development undertaking. If this is the case, then unlike the standard open-ended research grant, the project you undertake with a commercial sponsor must have defined and quantifiable goals. There will be financial consequences for not achieving these goals. Make sure that milestones are negotiated and included in the project and, if possible, tie in a related progress payment or penalty. Make sure the time frame (start and end) is clear and define the project very precisely. The latter is quite hard to do and will be related to your costs. Above all, make sure the project is a realistic one; nothing is worse than a project that fails, financially, politically or scientifically, for both licensee and licensor. You can always add bonus payments to the agreement for projects that could exceed expectations.

Defining the personnel

If you are undertaking a joint research project or indeed are performing contract research, then the personnel to carry out the project must be identified. This is particularly important with such agreements, since people move on or may prove to be incompetent. What are the consequences? The most obvious is that the project will not achieve its anticipated goals in the time frame that has been outlined and as such, there will be financial consequences. This applies to both yourself and the licensee. An important point to consider is the necessity for a full complement of skilled staff. The project milestones will have defined resources that are required in order to achieve them. One of the latter is the human resource and your ability to put together an effective team within in a short time period will be crucial. If the project begins without its full complement of resource, then it is likely to be subject to an overall delay and hence the risk of failure increases. You should also identify the consequences of a delay in recruitment and write a contingency into the agreement.

The licence agreement

There now follows a generic example of the type of agreement that you are likely to see as part of a licence. This in not intended to be an exhaustive list of points and others may occasionally occur, being germane to the circumstance in hand. No one can predict every eventuality and even if they could, a licence agreement would be extremely large indeed, with an extraordinary number of caveats. However, a good licensing lawyer can help minimise possible ambiguities and it is wise to employ the best you can afford. Just as a single base alteration in a DNA sequence can have a profound effect upon the activity of the resultant protein, so can the twist of a word result in a significant change of emphasis to an agreement, perhaps resulting in financial or operational consequences. Hence professional help, together with care and diligence on your part, is of profound importance. It should be noted that we are entering a time of significant change in the management of intellectual property in UK universities and you should ensure that the recommendations in the CVCP guidelines are incorporated into the agreement.

The agreement begins with the: (a) effective date from which the licence is active; and (b) name and registered address (or place of business) of the licencee, licensor and third party guarantor (if applicable). It continues by setting out the terms and conditions of the licence.

☛ Background information

- A brief description for the impetus of the agreement and a statement of what each party wishes to achieve. This can be useful in terms of future disputes and when there is a requirement to construe the object or purpose of the agreement.
- Nature of the licensable material and general comments as to who derived it, who supported the research, who administers the intellectual property.
- Policy statement of the licensor and general undertakings regarding the technology.
- Statement of the licensees aims and ability to carry out the terms of the licence.

☛ Definitions

- List of applicable IP rights such as patents (numbers, issue dates, jurisdiction, expiry dates, renewal dates), trademarks and goodwill, delineation of trade secrets, copyrights, design rights, etc.
- Of products, i.e. those materials that, in the absence of the licence, would infringe the patent claims. This can be quite extensive depending upon the technology. If we are dealing with a technology of wide applicability and generic use, then we may expect several categories, each of which must be separately defined. For example, end products may include formulations used for health care purposes, at

least one component of which is derived from the licensable material. It might also include reagents produced for research purposes only, but not those requiring further processing.

- Of financial terms such as net sales and first commercial sale (thus determining the point from which royalties become due).

Grant of licence

This to include a statement of the statutory rights, i.e. expressely what confidential information is to be transferred, who has permission to use it and an expression of the limitations of use (e.g. if some or all of it is transferable to a third party). It will also state whether the licence is to be exclusive, non-exclusive or sole. Exclusive agreements make exceptions of everybody except the licensee from participating. If the licensor is permitted to continue to operate and use the technology (perhaps develop new products) then the licence is referred to as a sole agreement. For example, two competitors may be involved as licensee and licensor, perhaps wishing to use the same technology for slightly different purposes. Non-exclusive licences are perhaps more general nowadays and are more applicable to biotechnology. You may have discovered a new process or enzymatic reaction and thought of a few applications, but this could be small when compared to the combined creativity of the world's scientists using your technology. So therefore, why confine yourself to such a limiting arrangement, whereby someone else may be free to exploit these other areas? Thus, as may now be obvious, the grant of an exclusive licence is a commercial decision based on market information. It gives the licensor greater control over the defined market and of course, this privileged position should not come cheap, i.e. it should be reflected in a higher royalty rate and/or upfront payment. The downside effects are of course the dependence upon a single licensee, first because they may not be as competent and efficient as the licensor expects and second because they may be inappropriate to exploit a hitherto unrealised profitable market segment. All agreements will contain a clause suggesting both parties use their 'best efforts' toward execution of the agreement but this is difficult to define in a reasonable manner and probably just serves to guard against a blocking tactic, i.e. prevention of a licensor exploiting a particular area. One must also consider that exclusive agreements are by their very nature anticompetitive and may well be challenged under the relevant countries legal procedures, e.g. by the UK Monopolies and Mergers Commission or by the antitrust laws in the US.

Exclusive licences may be possible where there are several clear technical applications of the technology or where it can be applied to a specific geographical region. For example, the inventor of a new technology (licensor) could agree not to exploit the invention in the territory requested by the licensee and not to sell similar licences to potential competitors. If definite technical applications can be identified, one could envisage several independent exclusive licences, without necessarily compromising new areas yet to be identified. The grey area is somewhat larger for the disbursement of technical information, several diverse items of which may be necessary in order to achieve the licensees objectives. In this case, it would be optional to package the information such that it could only be used for the purposes outlined in the licence. Not a trivial task. The licensee and licensor are allowed to agree that the licensor will not: (a) issue another licence to exploit in the territory of the licensee and vice versa; (b) actively market in each others territory; or (c) passively market in said territory for five years. Clearly depending upon the territorial scope of the patent, there may be exemptions. The next clause will specify the territory of the licence, e.g. a single country, multinational or part national or a specific listing of designated countries, all of which is dependent upon the territorial extent of the IP rights. There next follows the term of the agreement, which is usually within the time period of the life of the shortest intellectual property right, i.e. the patent. For example, with relevant patents A, B and C being necessary to carry out a function, there may be 5, 10 and 15 years left to run, respectively. The licence should then be limited to patent A since beyond that point it may be difficult to continue in operations. There may then follow a clause indicating an option to extend the licence and or an indication of reciprocal interests via cross licensing and the value placed upon them.

Payment schedules are then stated. The following options are common and will be a subject of negotiation; the route chosen will in part be dependent upon the goals of the licensee and licensor, together with their current cash position. For example, the degree of exclusivity offered will offset the value of the licence and the amount paid. Payment structures will reflect the degree of risk, the costs that are likely to be incurred during administration of the agreement and the cash flow. In general, for a small company it is better to get as much cash up front as possible. There should also be a clear statement of whether there is a provision for sublicensing the technology to another party, i.e. a company that is not affiliated to the licensee. In general, one should be careful with such clauses since they can result in a significant dilution in royalty income if poorly drafted. Furthermore, the mechanics of any sublicensing arrangement should be stated, together with the formal legal relationship of the sublicensee to the original licensor. There may also be a clause pertaining to payment of royalties for past infringments or usage of the technology prior to signing of the agreement.

What then, are the major determinants in licence fees? For the licensee there are three major factors that will be considered. (1) Does taking of a licence place restraints on future operations? For example, midway throughout the licence a new and competing technology may become available. Is the licensee free to examine and licence this technology? (2) How much will it cost to introduce the licensed material into the production schedule and into the market? (3) Does the licence place restraints on future competitiveness? Legislation generally favours competitive behaviour and frowns upon non-competitive actions. These are issues which should be considered at an early stage. The licensor has a similar set of considerations. (1) Is there any financial risk in offering a licence to a potential competitor? (2) To what extent are revenues (profits) diminished by relinquishing particular territories? (3) What is the likely value of new information developed by the licensee in using the licence? This could be in either technical information or commercial information, for example customer suggested improvements or market research. (4) What are the respective costs for establishing and servicing the licence?

There follows a brief description of payment structures (Fig. 6.1); in reality a mixture of these will often be used.

☞ *An upfront payment*

This is a lump sum whereby an amount of cash is received in return for the licence. This would be appropriate where an 'information' licence is being considered, especially for academic institutions that are generating new information in biotechnology related areas. Most biotechnology and pharmaceutical companies are looking for new technologies and desire to keep up with the latest developments. The half-life of much of this information is very short and even if you have something quite new, it is rarely long before other laboratories come up with something at best similar and at worse better. For example, a new procedure to purify a cytokine may only confer transitory benefit, but during that window, it could give a commercial concern a considerable advantage over the competition. Many such pieces of information are languishing in our universities, sometimes exchanged in return for consultancy to a few select individuals, but more often than not given away to shrewd commercial scientists on the telephone or by an invited seminar.

For payment, rather than being received in a single transaction, the 'information' licence may under some circumstances involve payments over several periods. This may not necessarily be dependent upon sales or products, but may be in return for a continued supply of information. This could be considered a long-term collaboration between the academic and commercial laboratory, for which information flows.

☞ *Returnable prepaid royalties*

This is equivalent to a cash advance, which is returnable as sales accrue, i.e. will be deducted in a predetermined way from royalties and a number of arrangements are possible.

☞ *A percentage of royalties upon sales*

A series of arguments indicates that determination of royalty rates should be on an empiracle basis, i.e. the going market rate for a particular piece of technology. This reflects both the differences in calculating royalty rates and the large activation energy that is required to calculate new ones. However, in biotechnology new precedents are being established and it is not sufficient to 'borrow' a

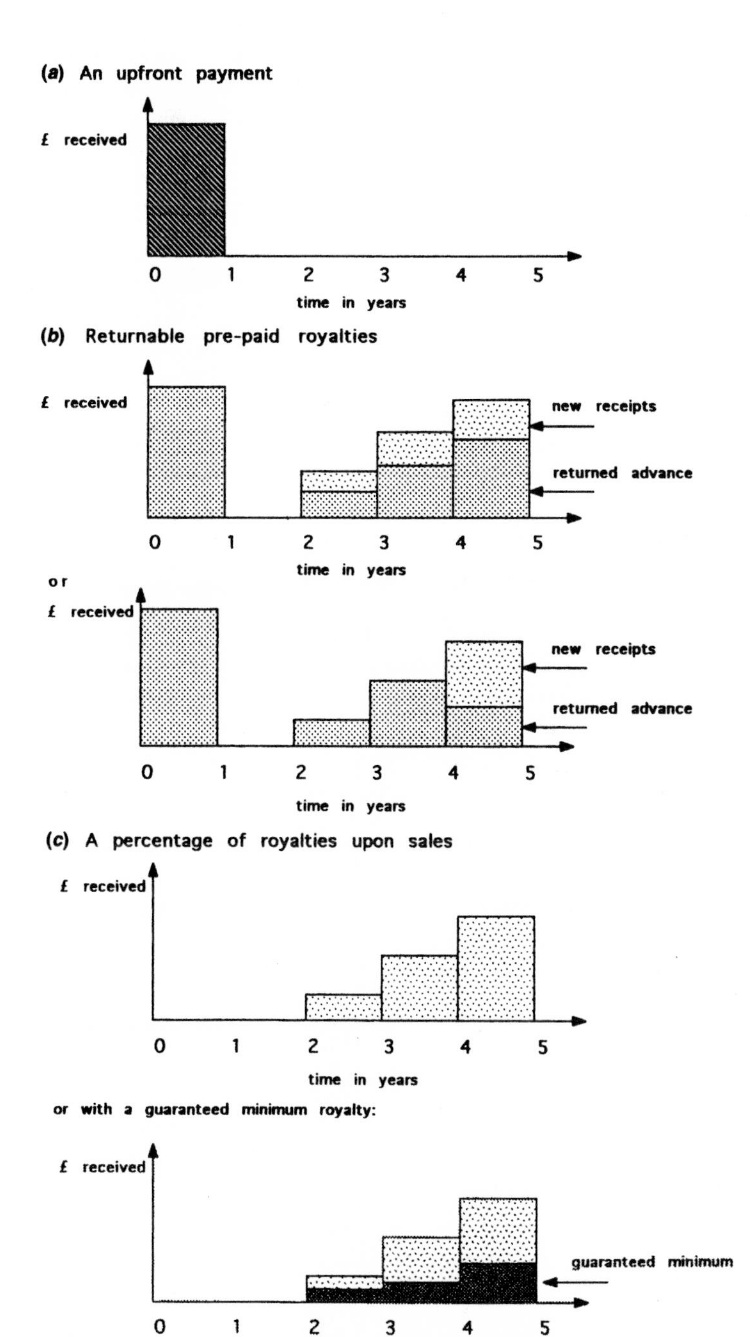

Fig. 6.1. Payment schedules.

royalty rate from a related field. However, the royalty rate is only meaningful as the amount that the prospective licensee is willing to pay, arrived at after much negotiation using your calculations as a guide. What is important though, is that you or your investors do not adjust the licence fees in order to recoup an investment in a particular technology. This could result in serious errors and may prejudice your negotiating position. The royalty will usually be expressed as either a percentage of net sales (or value of units produced) or in some cases a fixed monetary sum multiplied by the number made or sold. In any event, it is common practice to include a minimum royalty payment in conjunction with a 'best effort' clause under the operations section of the licence. This gives the licensee a minimum sales target, i.e. hopefully will encourage them to do the best they can to sell the product.

EU legislation does not allow an upper ceiling of royalty payments if the intention of the parties is to restrict product sales. However, some agreements may include clauses that either reduce or cease royalty payments beyond a certain volume. Since the licensees margins will increase beyond this point, there is an incentive to reach this target. Similarly, after passage of a pre-agreed time period the royalty obligations may be dropped, thus leaving in place a royalty-free licence. The royalty rate agreed should include a means of accounting for the inflation rate and the latter should be from a pre-agreed source and time. If calculating royalties from net sales, it is also important to define 'net sales' accurately. This will be the actual sales by the licensee and not the sales of a third party for which the licensed product is used. The net sales figure could be reduced by: (a) customer credit; (b) returned goods; (c) discounts for quantity purchased; (d) sales taxes; and (e) sales commission. If taken from the net sales figure by the licensee, this could result in a serious loss of royalty to the licensor.

The licensee wishes to maximise sales and since royalties to the licensor are likely to be a relatively small percentage of the whole, it is expected that the licensee will be more worried about their own revenues than that paid to the licensor. As a licensor, you will have little control over the selling price (at least in the UK and US) chosen by the licensee. It is possible to suggest a price (e.g. you make a kit for restriction digests, which is then licensed to a major distributor) but it must not appear to a court that there is a predetermined price. This is in an effort to prevent anticompetitive behaviour.

In this section of the licence there will be a indication of when the payments are due, e.g. monthly, quarterly. Each payment should include a **grace period** to allow preparation of the accounts. In addition, it is a good idea to tie this into the project management and reporting system. The currency of payment must then be stated together with a definition of the exchange rate to be used if royalties are expected from sales in other countries.

A statement of tax liability is then required. It is as well to remember that the tax laws of the country in question take precedence over any clause in your agreement. Many countries have arrangements for avoiding double income tax but the licensee and licensor must determine who is liable for national and local taxes where they exist. A suitable phrase would indicate that the licensee is responsible for all local and witholding taxes that may be due (e.g. VAT), which are not then to be deducted from the royalty payment that is due. These issues are important since such taxes could seriously affect the cash flow position of your enterprise. The royalty due to the licensor is considered to be income and is not generally considered to be a capital payment. The licensee is thus obliged to deduct basic rate tax in any royalty payment. Both the transfer of know-how and any sale of an intellectual property right will usually be considered to be a capital payment and hence they will be taxed as such. The licensors royalty is income in return for a service provided (i.e. is not a supply of goods) and as such UK based licensees are obliged to include a 17.5% provision for VAT. It can, however, be separately stated that the licencee may be liable for royalty derived VAT. Any income that you receive from foreign licensees is zero rated for VAT purposes.

A statement of the consequences of non-payment of royalties will also be necessary as will the responsibilities of any third party that may have been sublicenced. Since under the law of privity of contract, a third party cannot be either liable or benefit from a contract to which it is not a party, an express agreement may be necessary to create the liability.

A statement of procedures for policing the contract is necessary so that both parties can be sure that the correct royalties are being paid. Ideally, this could be linked to project management reports, including sales and production figures. Clauses attesting to the adequacy of the period audit and penalty clauses for mistakes should be included, preferably with a certificate from an independent auditor.

Stamp duty is not likely to be involved in a licence since there will usually not be a transfer of property rights. Similarly, stamp duty is not payable if the licence refers to a transfer of know-how. However, if the licence is exclusive and non-reversible, a court may consider that assignment has occurred. If necessary, a suitable clause may be inserted.

☛ Operations

The next section of the licence will contain clauses which attest to the management of the licence or how it will be done. This will indicate the respective obligations of the two parties and how they both perform their duties and interact between themselves and with others. Obligations on the licensee are likely to include the following points. (1) A determination of the commercial needs for the product in the market place and provision of reports as to performance. Hence there needs to be proof of quality control procedures carried out, so that the licensor can ensure the technology is performing adequately. Clearly, if you had sold the rights to a new enzyme (discovered by yourself) to a company who purified it to a low standard and marketed it, this could conceivably reflect back upon your laboratory at some point (e.g. customers referring to your scientific paper). (2) Collection of market information regarding the utility and size of sales, together with provision of information to the licensor regarding product improvements and new technological improvements. (3) Reassurance that there will be an adequate number of staff and other resources applied to the project. This refers back to the 'best effort' clauses, which is inevitably an unenforceable undertaking that the licensee will do its utmost to carry out the terms of the licence. It could also be specified that certain staff be placed in charge of the administration and execution of the agreement, and if these were moved to other projects without good reason, this could be accepted as a breach of contract. (4) The licensee will also be required to ensure that any local territorial rules and regulations are not contravened.

The obligations on the licensor are somewhat simpler and relate to verification of promises made. (1) The licensor must provide all of the information required for the execution of the licence and in a sensible form. (2) There may be obligations to provide back up services, whether this be documentation, materials or personnel. Any materials will be specified and of a defined quality. (3) There will be an obligation to maintain the intellectual property rights e.g. patent fees and to both enforce and protect the property rights, in the courts. In addition, there will usually be a clause to allow the licensee first option on maintenance of the patents, etc. if the licensor is for some reason either unable or unwilling to do so.

There follows a specification of the minimum performance expected of the licensee, usually expressed as a minimum sales volume. Since the price cannot be predetermined and the sales volume is unknown, the value of these royalties will not be clear until the first few payments are received and the trend established. This clause will be in conjunction with one suggesting that the licensee (and licensor) use their best efforts in execution of the licence.

There may then follow a non-competition clause suggesting that the parties do not compete in defined areas over a specified period of time.

A provision will be made for improvements to the technology in question, which to the licensee may include an obligation to advise of improvements, i.e. enhancement of the foreground IP position and a clause relating to the grant back of property rights, perhaps in exchange for a reduced royalty. This may include a clause that grants the ability to use the information (but not necessarily disclose it to a third party) once the agreement has terminated.

The next clauses pertain to confidentiality. In particular, the parties must take particular care to define the confidential information that is to be transferred, since the following classes will be exempt: (a) that independently developed by the licensee; (b) that information that is already in the public domain as of the date of the agreement or subsequently becomes so; and (c) that held by third parties. Confidential information is that which could damage the original source if revealed by a receiver and/or that which is stated as such upon transfer.

Following the confidentiality section will be clauses which pertain to the termination of the agreement and the settlement of disputes. It is an area that deserves considerable attention and sometimes does not receive it, negotiators being euphoric from the compensation discussions and not necessarily wishing to introduce negative influences. Nevertheless, one should attempt to write the divorce settlement before you undertake the marriage. This is not generally a politically wise move if you want to keep the trust of your future wife, but perhaps one that should be considered (in a dispassionate way of course!). Termination clauses should include the choice of country law and the appropriate legal forum. Standard reasons for breach of contract will include bankruptcy, a change in the specified personnel (which could affect confidentiality) and a change in ownership of the company.

Clauses will also ascertain the necessary notice period for termination and how this will be given, what processes there will be for arbitration and resolution of disputes, clauses relating to termination by force majeure (natural causes, i.e. outside of the control of the licensee or licensor) and any post-termination obligations, i.e. which clauses survive the termination date of the agreement.

To summarise, termination may be by: (a) natural expiry of term; (b) by giving notice without cause; or (c) giving notice with cause (failure to pay, failure to supply relevant information, lack of 'best effort', bankruptcy, a change in personnel or ownership, etc.). There will also usually be a specified period in which the opportunity is given to remedy the situation.

The final clauses will relate to execution of the document, such as registration, authority of the signatories, language used, the number of copies and where they are kept.

Types of licence

Licences are classified by the nature of the property rights transferred. However, the definitions are not always hard and fast; there are many shades of grey.

☛ An enumerated rights licence

This occurs when the property rights are listed and the numbers of the patents are given. One should be careful of licences that cover manufacture, use and sale. These three categories cover every aspect of the patent and you may not wish such an all encompassing arrangement. It is best to clearly define each of these and be sure of what you are trying to achieve, i.e. a licence to make and use a product is not a licence to sell it. Restriction upon the use of patents and products may only be valid if these uses are technically distinct, so again you should be clear on the latter points. The courts would regard it as anticompetitive to restrict the use of a patent or product unfairly.

Neither the confidentiality pertaining to the property right nor the return of information will be guaranteed by the licensee and so the licensor must attempt to make it so by inserting a relevant clause. The licensee must be concerned with any possible limitation on its own competitive policy as a result of taking the licence and with any limitations regarding their ability to challenge the nature of the property rights. Indeed a licensee may be developing similar technology yet still decide to take a licence. If the in-house technology proves to be superior, then is the licensee restricted from using it as a result of the licence? Finally, the licensee must be concerned with obligations that remain after expiry of the licence. On the other hand, the licensor will have obligations that relate to possible changes of ownership of the property rights and their maintenance. There may be additional undertakings not to exercise the property rights in territories that are not specifically licensed. This could affect the licensees ability to export.

☛ The field licence

This confers the right to operate in a precisely defined area and gives the licensee no access to the commercial or technical information of the licensor. This may be applicable when two laboratories have been concurrently developing a technology and one has won the race to the patent priority date. The payment to the licensor is thus not dependent upon the given property rights.

☛ The information licence

This relates to the transfer of operational information that need not necessarily be either confidential information nor a transfer of rights. For such a body of information, a royalty may be inappropriate and payment will be a single capital transfer without further obligation or consideration. The licensor will be expected to give relatively few guarantees and the price charged will be dependent upon the cost of preparing and transferring it.

☛ The technology licence

This is a fairly complex form of licence in which the licensor is under an obligation to transfer confidential information to be used for a specific purpose. The quality of information, supply of material and equipment and the detailed parameters or operational criteria must be precisely defined. In addition, the ability of either party to subsequently modify the technology may also be subject to precise controls, although this could have deleterious ramifications as both licensee and licensor open up a technology gap. One form of such licences are often called 'turn key' operations, whereby an operation is installed by the licensor at a pre-agreed location and the licensee does not have sufficient information to repeat the exercise. In this case the onus for success will be upon the licensor, and not only may **warranties** be required, but success fees for timely completion may be appropriate.

☛ Other arrangements

There are a plethora of other types of agreement, two of which are germane to our discussion. First you may consider an option agreement whereby you can either test a market or have your technology evaluated for commercial potential. In this way you could test a potential licensees ability to carry out the forms of a proposed licence. Obviously one should proceed with strict confidentiality and grant limited access to the relevant information.

Second, a joint venture may be appropriate. This embodies a new operation that is independent of the founding parties, despite dependence upon the licensors technology. One might consider a biotechnology start up funded as a joint venture by a university department and a pharmaceutical company, utilising the IP of the university and business skills of the commercial concern. A number of such relationships have been initiated, particularly in the US.

The effect of competition law upon your business

Competition laws exist to protect consumers from suppliers entering agreements between themselves to reduce competition in the market place. These laws prohibit certain kinds of agreement and the licensing of IP is one area where contravention may be possible. In general, these laws are designed to limit large companies whose trading can markedly affect the prevailing market conditions. However, if you have an innovative new product that creates a virtual monopoly by creating a new market, then competition law could be relevant to you. The two relevant pieces of legislation for UK companies are: (a) The Restrictive Trade Practices Act 1976; and (b) article 85 of the Treaty of Rome, the original agreement for the formation of the EEC. The former applies only to businesses operating in the UK and certain specified restriction relating to pricing of particular customers/markets. Most, but not all, IP licences are excluded from this legislation. Article 85 is of a more profound effect; towit it prevents agreements between EU companies that can prevent, distort or restrict competition within the single European market. The law applies to prevent agreements that: (a) attempt to partition or share markets; (b) fix prices; (c) prevent the licensee from challenging the licencors rights; and (d) prohibit exports. Exemptions can be granted by the European Commission if it can be demonstrated that although the agreement contains anticompetitive restrictions, these restrictions are required for commercial effect and there is an overall benefit to the consumer. If a patent licence is thought to fall within the prohibition, then individual exemptions need not be sought if they can be encompassed within a 'model; known as the block exemption. These cover commercial activities such as: (a) exclusive agency agreements; (b) patent and know how licensing; (c) exclusive purchasing agreements; and (d) research and development agreements.

7 Forming your own company

I must create a system or be enslaved by another man's.
William Blake
In *Jerusalem*, Chapter 1, 1815

Trading places

Before trading begins, it is important to consider the legal form under which you will conduct business. There are four types of legal entity that can be set up in the UK, the most relevant of which is a limited liability company (Ltd). We will briefly discuss the other forms (a sole trader, partnership and cooperatives) for completeness.

☛ The limited company

In this form of business your liability is limited to the amount you contribute by way of share capital. It is an independent legal entity and is separable from its owners, managers and employees. If the company fails therefore, any shareholders are not liable beyond the paid up value of their shares and creditors claims are only applicable to the assets of the company. In reality, this practice does not give enough security to lenders and you may be required to offer some guarantee (personal assets) to cover part of the loan. The limited liability company has a wider range of financing options when compared to other business forms. In particular, this can be by selling shares or creating a floating charge over its assets by selling debentures. Thus to summarise, the advantages of a limited liability company are: (a) it is a separate legal entity; (b) your personal assets will be protected (up to the value of your personal guarantees if you have given them); (c) there is a wider variety of security on offer; and (d) there are tax advantages upon which a specialist tax accountant will advise.

What then are the disadvantages? There is a greater degree of paperwork required in registration of the company (there are costs involved too!) and the accounts of the company must be audited by a chartered or certified accountant, which can be very expensive, especially in the early stages when you do not have a large turnover. However, under new legislation there are certain exemptions, details of which can be obtained from Companies House.

A number of other registered forms are possible, including a private limited company without share capital (limited only by guarantee), a private unlimited company with or without share capital, or a public limited company with share capital (a plc). However, limited liability companies with share capital are the most popular. If successful, a private company may obtain further capital by 'going public' and raising money on the stockmarket, as a plc.

☛ Other legal forms of business

Many choose to set up as a sole trader, where the business becomes one of your assets (e.g. your car) and subject to the provisions of the various Bankruptcy Acts, creditors may be able to acquire your assets in times of difficulty. The rules regarding record keeping are relatively informal (unless VAT registration is required) and professional audit is not necessary. However, whilst you will be your own boss, all of the capital must come from yourself or loans, which has inherent risk. Similarly, you may form a partnership with other sole traders as a means to acquiring more capital and skills. However, there are consequences should one of the partners make a mistake. A third form is the cooperative, in which the venture is owned by the employees. It is most unlikely that you would have control over profit maximization under such a format and it is unlikely to be appropriate to putting high technology into the market place.

Sources of company law

Parliament has enacted complex rules that regulate the activities of registered companies, embodied in a series of Acts of Parliament that are continuously being reviewed and updated to adjust for different trading conditions. The most current is the Companies Act 1989. Similarly, the European Union is attempting to harmonise company law in order to ensure that the trading conditions do not favour particular areas, although it should be noted that the legal systems in individual countries may still have their differences (England and Wales are treated as one).

Setting up your company

The process of setting up a company is actually relatively simple, Many people employ a solicitor for this process, but in reality this is not necessary. Companies actually come ready made in packages about the size of a telephone directory, and can be bought 'off the shelf' from a number of organisations that are set up for the purpose of company registration. These specialists are the first directors and company secretary and upon purchase, the rights are transferred to newly designated persons, the previous persons resign and register the fact with the Registrar of Companies. The cost of a UK company is between £90 and £220 depending upon the add on services that you require. Offshore companies are also available for slightly larger sums and could be advantageous from a taxation perspective. However, in the first instance you must choose an appropriate name for the company and check that it has not been used already.

Choosing a name

The name of your company should say nearly everything about it. It must convey the company direction with energy and enthusiasm whilst leaving space for expansion. There are many mutations of Gene, Bio, Tech, etc., some of which are still available. Others have looked to Latin or Greek translations or made variations in the name of the town which they operate. The Cam of Cambridge for example has been extensively used and whilst Cam-this or Cam-that may sound trite, it does an extremely good job at conveying excellence, since Cambridge University is internationally renowned as such. One problem however, is that such names do not easily permit expansion. The prefix Cam- would mean very little to a business trading anywhere else, except perhaps Massachusetts. One good tip is to include a reference to the technology you are using, e.g. cell, molecular, sensor, etc. since this will instantly convey a marketing message. If you are struggling to find a name, sit down with a blank sheet of paper and write down as many generic words as you can think of which relate to: (a) the technology; (b) the place and (c) the image which you wish to portray. Then mix them (and/or parts of them) to see what comes out. Be aware of little tricks like using abbreviated words, e.g. Nu for New, anagrams or pneumonics.

When you have chosen a suitable name, the next step is to phone Companies House to check if the name is already in use. This is commercially very important since some companies are very sensitive to use of names similar to their own, for obvious reasons. If you are unsure of the similarity to another name (even though the one you have chosen may not be identical to another) you should seek advice from either Companies House or a trade mark agent; the law in question is that of 'passing off'. A company can be bought off the shelf with ready-made names, in which case you could have your company formed within 1–2 hours. If you wish to change the name at a later date, there will be an additional fee (£50). Choosing a completely new name at the beginning will delay the process for approximately 1–2 weeks whilst registration of the name takes place, but the fees will be as before. If you do need to change the name, your company will have to hold an extraordinary general meeting and send both a copy of the minutes and the relevant documentation to Companies House.

Companies are registered with a given name providing that the name is not identical to or too similar to one already in use. In addition, there are some names that could be construed as something which the company is not and their use is restricted, for example, international, bank, group, chemist. Similarly, names that give the impression of being associated with local authorities or government departments are restricted, as are names that may be considered offensive or those which constitute a criminal offence. Further information can be obtained from the Companies House notes 2, 3, 4 and 11.

The memorandum and articles of association

In the package you receive, there will be several copies of the memorandum and article of association for the new company. The memorandum of association sets out the companies objects and constitution. A supplementary document, the articles of association sets out internal administrative matters, i.e. how the company is to operate in order to achieve its objectives. The memorandum must include: (a) the company name; (b) the country of registration; (c) the objects of the company; (d) a statement of limited liability; and (e) the amount of share capital and how it will be divided, e.g. a share capital of £100, divided into 100 shares of £1 each. The objects clause is to indicate the purpose for which the business has been established. The Companies Act 1985 permits a company 'to carry on business as a general commercial company', which is taken to mean that the company can carry out any business or trade and that the company may do anything which is incidental to the business or trade in question. Special (rather complex) rules apply if you wish to alter any of the details in the memorandum and the precise procedures should be discussed with a specialist in contract law. The articles of association are designed to regulate the internal affairs and management of the company and may be altered by unanimous agreement of the board or a special resolution. There is a model set of articles that has been proposed in the Companies Act 1985 called 'Table A'. The latter has been drafted by legal professionals and brings together over a century of experience in company operations. You may make alterations to this if you so wish and have these witnessed by the subscribers to the memorandum and witnesses thereto. Table A contains articles relating to: (a) shares and share capital (articles 2–35); (b) general meetings (36–63); (c) company secretary and directors (64–101); (d) accounts, dividends, etc. (102–110); and (e) notice of meetings (111-116). The company documentation (memorandum and articles of association) that arrives with an 'off the shelf' company may not be exactly what your purposes require and will probably need to be changed, which can be accomplished by holding a board meeting. It is common practice for shareholders to enter into a separate agreement to regulate some of the interactions between them and the manner in which the company is to be managed. It is a document that is used often to adjust the balance of power. Again, it is essential to seek professional opinion on contents of the shareholders agreement.

Share capital

Share capital is the fund to which any creditors can look for security with respect to payment of debts. This may be divided into issued, paid up, uncalled and reserve share capital. The issued capital is the nominal value of the shares that are actually issued and if they have not been completely paid for, gives rise to paid up share capital. Uncalled captial is the fraction of the issued capital that the company has not yet requested be paid and the reserve capital is that which is held in the event of liquidation. If the shares in the company are worth more than the nominal value, then the excess is referred to as the share premium, which must be retained in a separate account. There are four types of shares in general usage: (a) ordinary shares, which have no special rights or restrictions and can be subdivided into subclasses that may have a different value; (b) preference shares, which usually confer the right to preferentially receive dividends over other classes of shares; (c) cumulative preference shares, whereby the dividend can be carried forward to subsequent years; and (d) redeeemable shares, which are issued subject to the company having an option to buy them back on a fixed date or after a particular period. Additionally, the company may issue shares from its distributable reserves in proportion to existing shareholdings, which is referred to as a 'script' issue. Usually the issue of new shares must be first offered to existing shareholders on a pro rata basis, which is known as a pre-emption right. This may be negated by an appropriate clause in the articles or disapplied by the passing of a special resolution. The company has a variety of means by which it can alter its issued share capital, brief details of which are given in the Companies House note 30. The legislation and pitfalls that are associated with the holding of shares are legion and professional advice will always be necessary, particularly if you enter a relationship with a venture capitalist, if you decide to instigate an employee share option scheme, if you wish to transfer shares, etc.

Disclosure, the annual returns and other issues

There are five methods by which company information may be disclosed and there is a statutory need to do so for the first four of the following: (a) Information required by the Company Registrar; (b) information available at the companies registered offices; (c) notices in the *London Gazette*, a daily publication from HMSO that is used for notification of company changes; (d) publication in business documents; and (e) via the media.

In your business documents, disclosure of the name of your company is very important since as it is an artificial entity, the name is the sole means by which it can be recognised. Thus the Companies Act 1985 dictates that every office of the company should have its name displayed outside and failure to do so can result in a fine. This is particularly relevant if it becomes necessary to serve documents at the registered address. Similarly, the company name must be legibly mentioned on all business letters, notices, official publications, bills of exchange, cheques, invoices, receipts, etc. In all cases, limited (or Ltd) must also be included if applicable. All business letters and order forms must also include the place of registration, the company registration number and the address of the registered office.

Every year certain information must be passed to the registrar on a 'return' date. If you have not already got a yearly return date, this is the anniversary of the date of incorporation and it is an offence not to do so within 28 days of this date; all directors and the company secretary could be held responsible. The registrar will send a partially completed shuttle form (363) before the return date and all you need do is register changes. There is a £25 annual registration fee. Useful booklets that delineate the disclosure requirements and director's responsibilities can be obtained from Companies House (Companies House notes 9 and 15). In particular, directors are responsible for compliance with filing obligations and not the professional advisor to whom the task has been delegated. It is therefore wise to ensure that the requirements have been met.

Also in the package will be forms with which to record the identity of the directors/officers and exactly who owns the shares. There will be a book for recording the minutes of board meetings, a certificate of incorporation showing the company registration number and the company seal. The company will need to have a registered office somewhere and it is often the address of your solicitor or accountant. Other premises are of course possible. Once the name has been chosen and the company registered and incorporated, the next step is to open a 'small business' bank account. Most of the major clearing banks produce a small business 'start-up guide'. Whilst these are designed to be applicable to any business, there will be some useful tips for your high technology start up. It is worth talking to the small business advisor and seeing which deal is the best. Banks are always keen to take on new businesses if they think they are going to eventually be successful and meanwhile may make some money on loan and overdraft interest. Upon starting the business, you should first inform the tax office. Ring the Inland Revenue (locally) and ask for booklet IR 28 'Starting in Business'. This has a form 41G in the centre, which is simple to fill in. The taxes you need to consider are value added tax (VAT), pay as you earn (PAYE) or income tax, corporation tax, capital gains tax, stamp duty, customs and excise duty, national insurance and eventually, inheritance tax. Dealing with tax is unwise if you are not qualified to do so and is a complex matter. First, it is important that you get it right, since if you do not the Inland Revenue will and second, there are legitimate tax savings that can be made on your behalf by someone skilled in the area. The most important thing for you to ensure is that proper records are maintained and that you plan ahead. Inept management of tax will have a severe effect upon your cash flow.

The next step will be to register the company for VAT with HM Customs and Excise such that you can recover VAT on initial expenditure. For this purpose it is important to keep the receipts and invoices for each cheque issued or received. The registration forms may be obtained from your local VAT office (consult the telephone book). Your VAT office may decide not to register your company until you begin trading, which has the disadvantage that you will never be able to recover VAT in expenditure incurred during the start-up phase when the availability of cash is at a premium. This action is to stop individuals from avoiding VAT by purchasing goods through a company, when the goods will ultimately be for personal use, e.g. hobbies. If this happens you must telephone the VAT office and convince them of your business plan and intention to begin trading as soon as possible. VAT registration is relevant to companies with a turnover of greater that £45 000 per annum (raised from £37 500 in the December 1993 Budget). VAT is chargeable to your customers on goods and services and the amounts must be sent to Customs and Excise on a monthly or quarterly basis. You will be charged VAT by your suppliers for materials, telephone bills, legal and accountancy fees, etc., which can be claimed back. In order to simplify the process you deduct one from the other and pay or claim the difference. When you start up, it is likely that you will be trading cash negative, i.e. costs will exceed sales and so you must be in a position to reclaim the VAT. Useful documentation is available from your local HM Customs and Excise office. In addition, your new company must register for PAYE with the Inland Revenue, because any salary you take from the company will be subject to income tax and as an employee of the company this will be paid by PAYE. This is of course essential if you take on more employees. The local tax office will provide a booklet, 'Thinking of taking someone on?' number IR53, which will help in this process. Deduction and remission of income tax and national insurance contributions (which includes employers national insurance contributions) is a legal requirement if the employees earn over certain amounts and are considered to be in regular employment. Upon taking someone on, the tax office must be informed and in turn will inform you of necessary deductions. You must record employees earnings and deductions and submit this to the tax office each year. As for national insurance, the employer and employee each pay part of the contribution. Information on how to do this may be obtained from your accountant or the Department of Social Security. As a director of your own company, national insurance will be assessed on your salary and the company will also be required to pay an additional contribution. If you are an academic and still wish to retain your academic post, this will be carried out for you by your academic employing institution. The merging of the two requires proper financial advice. Some of the other forms of tax will be dealt with later but the key point is, get expert help from the earliest stages.

Accounting principles for your business: legal aspects

Once a year, the directors of a company must present an account of how they are running the company. There are two forms in which this information must be presented, a balance sheet and a profit and loss account. Both documents may appear simple at face value but are technically difficult to prepare and this must be done by a qualified accountant, the rules being determined by a series of rules known as the Statements of Standard Accounting Practice (SSAP). These accounts will also be used by those outside the company to assess you for financial arrangements, determine your credit status and work out taxes due. The registrar of companies holds these accounts and they can be inspected for a small administration fee. In the case of private companies, these may be lodged as summaries, although full details must be kept for a minimum of three years. Accounting records must contain: (a) a record of liabilities and assets; (b) a cash in – cash out book; (c) a statement of stock in hand, where this stock originally came from and of all goods bought and sold. All parties involved must be identifiable. The accounts must be approved by the board of the company and signed and dated on their behalf by one of the directors. For each financial year, the directors must prepare a review of the business during that period and also indicate the disbursement of any profits. Again, this must be approved by the board, signed and dated. An auditors report on the accounts and directors review must accompany these documents and be sent to everyone who is entitled to attend general meetings or who hold company debentures. Not surprisingly, there are also time limits for the presentation of this information which must be within 10 months of the accounting period for a private limited company. There are certain prescribed formats for the balance sheet (2 choices) and for the profit and loss account (4 choices), and once chosen these formats should not be changed unless there are special reasons. An accounting reference date must be chosen between 6 and 18 months after incorporation, using form 224 that can be obtained from Companies House. The first accounts must be delivered at the sooner of 10 months from the accounting reference date or 22 months from incorporation, otherwise a fine will be payable. Companies House must receive, within the alloted time period, a profit and loss account, a balance sheet and an auditors report if you have a company that is not eligible for exemption. Late filing of accounts may incur a fine and a penalty; there are generally no acceptable mitigating circumstances (Companies House note 22).

The directors report

Again, the Companies Act 1985 is specific about the content of the directors report. If you have a small company then you need not send a report to the registrar, since there is little point in preparing a report that only yourselves may see. It may be a good idea to prepare one for your own purposes however, as a review of what you have achieved and where you are going. The formal content of this report is indicated (Fig. 7.1).

1. Application of profits
2. Significant changes in the fixed assets
3. The names of those who held directorships during the financial year and the extent of their shareholding
4. Research and development activities
5. Charitable and political donations
6. Facilities for disabled persons (>250 employees)
7. Facilities for health and safety at work
8. A statement as to efforts to inform and consult employees (>250)

Fig. 7.1. Contents of the directors report.

What are your obligations as a director of your own company?

Every private company must have at least one director and a company secretary who is not also a director, or officers as indicated by Table A of the articles of association, if indeed Table A is used. All organisations require a governing body that determines its objectives and strategy, and who take responsibility for its activities. This function is carried out by the board of directors who will be expected to determine strategy (in respect of business activities, long-term objectives and balancing the interests of the company stakeholders – shareholders, employees, creditors, etc.), appoint top management, monitor progress and provide accounting information.

The board should ideally operate as a committee to determine the collective view. As such, directors are entitled to information that they require to perform their function and are entitled to rely on the information provided to them. Notice of meetings and the constitution of a quorum will be delineated in the articles although there is actually no legal requirement for the provision of an agenda. It is, however, customary for the company secretary and managing director to provide an agenda in order to increase the efficiency of the meeting. Most board papers include minutes of the previous meeting, which members will be asked to agree as a true record of the proceedings. It is preferable if the minutes record decisions rather than document discussions. There are special provisions for the holding of both directors and shareholders meetings and it is recommended that a suitable text be referenced (Appendix 4).

The board papers should include management accounts and relevant accounting statistics, together with the expected profits of the company. It is vital to closely watch the cash position, not least because of the effects of insolvency and the liabilities that derive therefrom. The board will also be reviewing new projects, and the progress of older ones. Full budgetary approval needs be given before projects are initiated and it is essential that the board behaves in a positive and proactive manner, i.e. does not just approve a retrospective report. The board is led by a chairman, who is also expected to be the company icon, presenting the company to the community. The role of the chairman is to ensure that adequate time is allocated to each item on the agenda, to sum up discussions and to try to ensure harmonious functioning of the board. The other key appointment to the board is that of the managing director who has personal responsibility for the success of the companies operations. Occasionally the two roles are combined, often using the title of 'Chief Executive' which is perhaps more common in the US. This is not usually a good idea since it could result in insufficient outward vision. In a small start-up company, the ideal structure would involve the inventor(s) (e.g. as scientific director(s)), business managers and an external chairman from either the academic or business community. Executive directors may devote the entirety of their working time to the company in question, often have a significant financial interest in the well being of the company and may be employees. In general they are remunerated separately from those fees that are applicable to board duties. A non-executive director receives a fee for a portion of the time spent in pursuing the companies activities, although under company law the two types of directorship are treated as being equal. It is advantageous for the company to appoint such external directors, in order to provide a wider perspective and different set of experiences from which the business should benefit.

The appointment of the first company directors has already been discussed in the section describing registration of the company. Table A allows appointment of new directors by an ordinary resolution at a board meeting. However, this person must agree to the appointment and not be insolvent or disqualified from acting as director by a court. There are a number of ways in which directorships may be terminated such as dismissal, retirement by rotation, assignment, resignation and other conditions according to the articles of association.

What then are the proceedings directors must execute, what are their powers and what should they be paid? Needless to say, there have been many court rulings on these subjects and the rules of acceptable behaviour are very complex. Only a few of the major principles can be given here and you are referred to an introductory text on company law for further details. Directors may perform their duties as they see necessary, i.e. put in as much or as little as is seen fit; if Table A is used, then a director may call a meeting at any time, you do not need to give notice to a director who is outside the UK, the majority decides upon issues during the meeting and the chairman has a casting vote. The second point is often modified in the articles. Quorum must be fixed by the directors (default is 2) and if a meeting is inquorate, decisions are not binding. Minutes of meetings must be kept to register the appointment of company officers, the names of those present and the proceedings of the meetings, although a decision is not invalidated if you fail to minute it.

The directors may appoint a chairman or remove him from office at any point. If the chairman is more than five minutes late, another may be appointed for that meeting only. The appointment is important because of the casting vote given in Table A of the articles of association. Powers may be delegated to a managing director (or chief executive officer) appointed by the board on conditions of their choosing. Unlike other directors, the managing director is not subject to retirement by rotation.

The directors have general power to manage the business of the company as a going concern and subject to the memorandum and articles of association. Additional specific powers may also be granted. A director has no implied right to be paid for services rendered unless it has been expressly permitted. In reality, provision for payment of fees is usually made in Table A, as is payment of legitimate expenses. The company accounts need not necessarily detail how much each director receives, but the aggregate for all directors must be shown. This is often referred to as the directors emoluments. If your company make payments into a pension scheme on behalf of directors, then this must also be stated. In exchange, directors owe particular duties to their companies, encompassed in a fiduciary duty (trust and confidence), conflict of interest and a duty of care. Each will be examined in turn.

☛ Fiduciary Duty

In the context of the company this means that directors must not behave in a manner that is detrimental to that company. That is, they must act in the interests of the company and not for any parallel purpose that may benefit them in some other way. Three tests of abuse of power are likely to be applied by the courts: (a) has transaction been performed to the benefit of the company and the interests of promoting prosperity; (b) is it *bona fide*; and (c) is it consistent with the normal business of the company, i.e. justifiable in the normal course of business authority?

☛ Conflict of interest

Directors have a duty to the company in that they are not allowed to make an unauthorised profit or should not get in such a position as to have conflict between their duty to the company and duty to another. Inevitably, persons will often have interests in more than one company and if a proposed contract brings these businesses together then the director has a statutory obligation to declare this interest. If a director was offered an opportunity in a business area similar to that of his company then following it up would be a conflict of interest. For example, if you were a director of a company that sold restriction enzymes then there would be a conflict of interest if someone approached you (knowing you had expertise in this area) and you helped commercialise a new enzyme by a route that did not involve your company. In many companies, Table A is now modified to exclude rules regarding conflict of interest and profit making, provided the interest is disclosed.

☛ Duty of care

Directors of companies must not be overtly negligent in the execution of their duties. They are expected to behave with such care as is reasonable for someone of their knowledge and experience, which essentially means that errors of judgement, otherwise in good faith, do not leave the director liable.

Shareholding as a director

If you are a director of your company, you will no doubt own some shares in it. Indeed, the articles may dictate that qualification shares are held. Directors will breach their duty to the company if they use their 'inside' knowledge to deal in the company shares. This 'insider dealing' is regulated by the 1985 Companies Security Act, whereby it is illegal to deal in shares of a company if the individual has access to unpublished confidential sensitive information.

In addition the Insolvency Act 1986 is also relevant to your duties as director, under which you are obliged to show the care and skill expected of a 'reasonable director'. If you do not do so, then if the company becomes insolvent you could become liable for any company debts. The 'reasonable director' will probably be determined on an individual basis by the courts.

Information for the Registrar of Companies

The following documents must be sited with the Registrar of Companies in Companies House: (a) appointment/resignation of directors/company secretary; (b) the registered office; (c) the memoranda and articles of association; (d) resolutions from general shareholder meetings; (e) resolutions which increase share capital; (f) share allotment; (g) the accounting reference date; (h) the annual return; (i) the accounts and directors report; and (j) winding up. There are a number of other requirements and you are advised to seek legal advice for each special circumstance, especially if you are not experienced as a company secretary. There are also a number of documents that must be available for public inspection: (a) a register of company members and debenture holders; (b) a register of directors/secretaries and their interests in the company shares; and (c) a register of charges (Companies House note 23).

Individual responsibilities

Nearly anyone can be a director of a company although sometimes there are restrictions written into the articles. Every non-public company must have at least one director. All directors of the company should have a service contract that must be available for inspection at the registered office or principal place of business. If you carry out a function for the business that is in excess of your board duties, then you should ensure that you have a written contract of employment to cover pay, working hours, duties, etc., indeed all of the issues which would normally be found in a contract of employment. The purpose of the board is to have a full and open discussion regarding the commercial direction of the company and from time to time disagreements will occur, for all the usual reasons. These should be duly minuted and if this is not sufficient or the arguments are continuous, the director in question should consider resigning. Once resigned (oral or written), withdrawal is not possible and the court will protect the company from unauthorised use of information obtained by the director while in office. However, if the disagreement is over an issue that the director considers to be either unlawful or unethical, then the director would be expected to remain in position and take steps (e.g. use of professional advisors) to effect a remedy.

The responsibility for the companies compliance with accounting practice rests with the directors, which may perhaps seem to be rather a chore to those of a scientific background. The directors duties are to: (a) ensure that proper accounting records are available; (b) to approve annual accounts which give an accurate representation of the financial position, i.e. a profit and loss account and a balance sheet; (c) send copies to those who should receive them; and (d) make sure that they are received by the Registrar of Companies.

Directors do not have any automatic right to be remunerated for their services and this must be provided for in the articles of the company, usually with phrases allowing such fees to be determined by the board. Directors fees are subject to income tax, national insurance and PAYE.

Insurance for directors

The legal position of a director brings with it an onerous series of responsibilities and duties, for which there is a very real and personal risk of breach, however unwitting. One might consider that directors have the same limited liability as shareholders, but this is in fact not the case, even though they may be shareholders as well. Indeed, in cases of insolvency there may be unlimited personal liability. If you are considering directorship, you would be well advised to consider insurance to cover against trading while insolvent, criminal prosecution, lible, false accounting, etc. Remember, such charges could be proffered against the business for which you are a director and you could be held equally responsible even though the alleged offence was nothing to do with you.

The Institute of Directors

There is a body that represents the interests of directors and provides an excellent forum for discussion of the legislation surrounding the company. The Institute of Directors can be joined for a modest fee upon recommendation by an existing member and you should consider joining if it is at all possible. It provides premises in central London for meetings, as well as producing a series of useful publications and running relevant courses. In addition, there is a business information and directors advisory service, together with the option of preferential rates for obtaining directors professional liability insurance.

Other functionaries in the company

Besides the managing director, chairman and directors of a company there are also necessary officers of the company, which includes the company secretary, the alternate director, managers and the auditors. The duties of these officers are outlined below.

☛ The company secretary

For a private company, anyone can be appointed secretary, provided they agree. This can be one of the directors or your solicitor or accountant. However, for public companies appropriate qualifications are required (Companies House note 16). The main responsibility is to ensure that the administrative duties which relate to the company are fulfilled, although these duties are not specified in the Companies Act. In particular, this means ensuring that the Registrar of Companies has all relevant information in a timely fashion. A company secretary is not necessarily involved in the management of the company *per se* but may be so in a separate capacity (e.g. as a scientific director). The secretary can lighten the burden of knowing how to comply with the detailed provisions of company law and may give independent advice on matters for which they are responsible. If a secretary is also a director of the company then they could become compromised with respect to their statutory duties and so this should, in general, be avoided. Duties of the secretary in a small company could include:

1. Maintenance of the statutory registers, i.e. the register of members, charges, directors and secretaries and directors interests.
2. Prompt filing of statutory forms.

3 Informing members and auditors of meetings within the acceptable time frame and communicating resolutions and agreements to Companies House. A resolution is a decision or agreement made by the members of the company that binds the company until it is superceded by another resolution. There are at least seven types of resolution, each for different purposes (Companies House note 26).
4 Supply of accounts to members to those entitled to them.
5 Keeping if the minutes of directors meetings. Details are in Companies House note 16.

☛ The alternate director

In the temporary absence of a particular director, an alternate director is empowered to carry out the duties of a director, but usually only at board meetings. Since the alternate is responsible for his/her actions when present at board meetings, the alternate must be registered at Companies House. Such a person will have both a personal vote and a vote empowered by the absent director. The articles of the company determine the power of the alternate who is usually appointed by a particular director, the latter having the discretion to remove them. The alternate director is generally not remunerated.

☛ Managers

Anyone who has a supervisory category or works in general administration according to the general policies of the company, is considered to be a manager. This is distinct from the responsibilities of managing director or a general manager. Such persons need not be a director or indeed be under specific board instructions but, have the same fiduciary duties as a director.

Taxation

When you have a commercial business, taxation issues become one of the most important factors to be considered and you should seek advice from your accountant as soon as possible. There are five types of tax/duties that affect IP rights; income, capital gains, value added and corporation tax, together with stamp duty. Different IP rights are subject to different tax regimens, e.g. patent licence income is subject to deduction at source whilst income payable to the holder of a copyright licence is not. The tax treatment of licence income depends upon: (a) the type of right that has been licensed; (b) the location of this right; (c) exclusivity/non-exclusivity; (d) revocability/irrevocability; and (e) in what capacity the innovator has received the income (e.g. as a limited company).

If the whole of an asset is disposed of, then it will be considered to be a capital gain and not earned income, subject to specific exceptions. If the seller retains some part of the IP, then the tax position is likely to relate to income, no matter how paid. In general patents, know how and copyright are not considered capital gains, especially if the sale relates to the sellers normal profession. Indeed, there are provisions within the relevant act to spread capital sums received forward for six years so that the whole sum is not taxable in the year of receipt. Sales of IP rights are also considered taxable supplies and if turnover exceeds a certain threshold, then value added tax (VAT) will be payable. In general, licences of IP rights are not subject to stamp duty. In dealing with the taxation position, it is essential to seek advice from an accountant. This will save you money and ensure that the relevant legislation is adhered to.

Insurance for the business

The option for IP insurance has been considered previously and is essentially independent of your business start up. If you do take this route then both the investors and yourself may require the additional security afforded by IP protection. In addition, there are various forms of insurance that should be considered before you either begin trading in earnest or take on any employees. Your insurance policy is the servant of the legal arrangements. It is designed to protect you from any legal gaps in agreements (contracts, etc.) that you may have entered into. No agreement can cover every eventuality and there will always be loopholes, which may be revealed at a later point. The simplest type of policy is the 'office combined' that will cost between £250–500 for minimal cover, to be taken when you begin trading. This should be adequate to cover property, pecuniary and liability claims. Of the latter, there are three levels; employers, public and product liability. The first is the only compulsory insurance and covers against personal injury to employees, but not damage to their property. Public liability insurance covers against third party liability, i.e. damage to someone elses property, personal injury to a third party, etc. The indemnity will probably be limited, with the minimum advisable being £1m. In practice it may be more sensible to take out up to £5m of public liability cover. The product liability may not be immediately relevant to your company; you must assess objectively the risk of it harming a third party. Specifically, this insurance covers damage to the property or person of a third party that is caused by the product, i.e. the product must directly cause physical loss. If you intend to sell your biotechnolgy product directly, then you will need this type of insurance. It is more likely that you will license a manufacturer who will then interface directly with the client. If you can negotiate a 'hold harmless' agreement whereby you cannot be held responsible for negligence, then your need for insurance will be reduced. To an extent, negligence will be covered by professional indemnity (PI) insurance, which should also be carefuly considered. PI insurance covers against any negligent act, error or omission on the part of any employee of the company. You should be considering cover in excess of

£1m that will probably cost around £5 000 and as before, if you can negotiate a hold harmless clause then this should reduce the extent of your PI cover. You should take out this policy when the information from the company first becomes disseminated, i.e. when you have an income stream, since you cannot forecast when a claim might be made. Depending upon the nature of the business you could require medical PI, but if not, there will be a wider range of underwriters available to you and your insurance agent may be able to find you a cheaper policy. It is important to remember that PI policies should evolve with your business, since as it grows the chances of a mistake increase. At the beginning of the enterprise it is likely that you will be unique; perhaps there is a new invention or you can provide a service that cannot easily be replicated. However, as your business develops, so a degree of delegation will be required and the normal tenets of supervision begin to breakdown, thus increasing the risk of mistakes. Obviously an adequate training programme could help circumvent this and reduce the size of your PI premium. You should review your policies on an annual basis to ensure they are balanced to the perceived risk.

The business plan

Each and every business requires a plan. This is particularly important for new biotechnology start ups as a wide range of skills will be required to turn the dreams into reality. In general, many scientists have not yet acquired the necessary skills in finance, marketing, manufacturing, etc. that will be required for the adequate development of the business and will, of necessity, have to draw upon the expertise of others. The purpose of this section is to introduce these functions so that you have an appreciation of these disciplines (including a selection of useful buzzwords!) and to help you in production of the first business plan. This will determine the potential the viability of the venture and will decide the route taken. It is important to remember that preparation of the plan is a reiterative process and will go through several revisions, as more information is collected and the plan modified in light of this. Do not expect to get it right first time and do not expect it to read like a scientific paper. Whilst a degree of complexity will be necessary to show that you understand your own plan, there is little point expanding on every little nuance or calculating figures to several decimal places. The tide of events is rarely as we predict or necessarily would like, so it is important not only to build in flexibility, but show that you can be flexible in response to changing circumstance.

☛ Why prepare a business plan?

First and foremost, writing a business plan will help you focus your ideas. Hopefully, this crystallisation process will lead to you having a significant degree of belief in your proposal and will take you on from a hazy idea to a project worth fighting for. It is this commitment that will be expected by prospective business associates. In addition, you can make mistakes on paper that are likely to be less expensive than those in the market place. Second, the business plan gives you a *modus operandi;* a structure for future management and future directions, with inbuilt fail-safe mechanisms for when the going gets tough. Third, it will provide the documentation necessary for attracting finance, whether this be via a bank, venture capitalist or other route. Such financiers expect the information presented in a particular way, and since you will be in competition with several other proposals, many of which could be outside your sector, the plan must be one of the best. This means that your meticulous planning, concise presentation and tight objectives will all be assessed as indicators of your ability to carry the project through to a successful commercial conclusion. In addition, a well-written plan will communicate your ideas to your professional advisors, i.e. solicitors, patent agents, accountants. This means they will be better able to assess the needs of the business and thus decide how to help you best. Fourth, the business plan gives you a method for determining your progress over a given time period. Thus if you either exceed your expectations or fail to reach them, the underlying reasons can be identified and adjustments made. For example, as you move into a new market so information will accrue that will necessitate adjustment of the business plan.

☛ To whom is the plan being presented?

For whom should the business plan be written? The first draft should be for yourself and/or your co-inventors in the technology. This will enable you to be confident in the route you are taking and will engender the commitment of those concerned. It will decide the scale of the potential business and the type of product, which then predisposes the marketing strategy. It will identify the potential customers, which again will suggest how best to proceed with the marketing. Most importantly however, it will help decide the amount of finance needed and the risk element involved. This in itself helps determine who (if anyone) to approach for capital investment.

The second plan will incorporate many aspects of the first, but should now be written with specific people in mind, e.g. the bank, venture capital institutions, etc. The first thing to do is sit down with a blank sheet of paper and write down the main points that describe the business idea.

- Define the service or product and the state of each (this could be research, development, ready for market).
- Consider the IPR position of both yourself and competitors. Does anyone else have a competing product?
- Describe yourself and why you believe you should be involved. Decide what you want out of the whole process.
- Describe who else is involved. What skills do they bring and what are their aims

and objectives? Do not worry if you and your collaborators are discordant at this point. It is better to be honest!

A useful framework under each of these is to perform an abbreviated form of a 'SWOT' analysis. SWOT stands for strengths, weaknesses, opportunities and threats and is often used as a management consulting tool to appraise a business venture. In contrast to conventional wisdom it is better to analyse your business from the reverse direction, i.e. do a TOWS analysis. This then represents more faithfully the actual thought processes that are likely to occur and the analysis is likely to be more accurate. The analysis is relatively simple and requires a creative look at the business, yourself and the environment. Do not worry if you cannot fill all categories or if the threats outweigh the opportunities, just try to be as objective as you can. A second technique, which should be seen as complimentary, is a 'PEST' analysis where this acronym stands for political, environmental, social and technical. Again, you can analyse your potential new business the light of these factors. For example, a transgenic animal business may not present too many technical difficulties, but is likely to score poorly on the social scale. AIDS research is likely to be high on the political agenda, but technically difficult. All of this helps you to understand the limitations of your business and will help you launch off in the right direction. What can appear as insurmountable problems can often disappear upon further development. However, the converse may also apply. It is essential that you are truly objective in this analysis, so that unforseen circumstances are minimised and you play to your strengths.

You must write the business plan yourself, thus effectively communicating your beliefs and enthusiasm that are, lets face it, critical to the success of the venture. You will be expected to justify the detailed content of the plan to potential financiers, which is extremely difficult if someone else has prepared the plan. Outside advisors should, however, be sought to help in the preparation of the plan and whilst it is unlikely that they will have seen a plan identical to it, they will have seen many like it and will have a good idea of the criteria for success. An accountant will help to: (a) establish accounting policies for the business; (b) review the financial projections; and (c) advise upon taxation. Your solicitor and business development manager will: (a) act as a source of advice regarding the legal issues associated with business start up; (b) draft and assist in contract negotiation; and your patent agent will (c) manage your intellectual property. Marketing advice will also be valuable although it may be perceived as imprecise by many scientists. We will be considering this subject in some detail but for now, remember that market research can indicate: (a) the total size of the business; (b) the degree of growth potential for both the company and the whole industry sector; (c) market acceptance of your new product; (d) potential competition; and (e) how to formulate a coherent marketing strategy. The plan should err on the side of brevity for the venture capital industry, who will employ their own industry analysts. Thus a shorter, more focused plan is likely to capture the imagination.

For the banks, err on the side of simplicity. For an average view on the length of the plan, a request for £1m will require a plan of around 15 pages.

Forecasting the financial results

This is probably the most difficult step for those of a scientific disposition and it requires considered attention. Most of us cannot predict what we will be doing next year, let alone in five years time. Why then do business plans require financial forecasts for three to five years hence? The first reason is to provide a means to estimate the company performance and second, it delineates the company strategy in financial terms.

☛ A performance measure

If you were in the position of the investor, how would you decide if the business has performed well, given all of the variables that could make an impact? In scientific terms it is equivalent to deciding the research direction in three to five years time, having regard to the results of the experiments you are conducting today. Since the whole point of making the investment is to make money, these criteria must be financial and consequently they must be deeply planned and considered by the entrepreneur. In biotechnology the short-term forecasts are often exceedingly difficult due to the volatility of the markets; long-term projections are quite likely to be very inaccurate. They should still be included, but the sources of volatility should be analysed, perhaps by providing a **sensitivity analysis**. In this way contingency plans can be made for uncertainty and it is as well to have these circumstances documented at an early stage. The other characteristic of biotechnology start up companies is that they tend to run cash negative for several years, having to go through several rounds of refinancing before a product enters the market, especially if the aim is to take a high margin pharmaceutical product through clinical trials. This means that the financial performance may be unrelated to the value of the business, since research and development may have gone to plan, there may be a series of compounds in phase I and phase II trials, but as yet the products have not entered the market and so cash flow has not yet been generated. Furthermore, the company may have put considerable resources into other assets such as intellectual property and business agreements. These assets must be included in the equation for measuring performance. The business may also need to take strategic decisions on the basis of their research (short-term) achievements and you should try to produce a financial model to illustrate the consequences of each of the options.

☛ **Delineation of the company strategy in financial terms.**

There are several areas in which to pay attention. First, the rate of growth relative to other companies in a similar sector and the sensitivity to market changes. Second, the impact of fixed costs upon cash flow, since the latter are incurred in a quantal fashion. Third, other influences upon cash flow, such as the timing of refinancing, product launch and a host of others. It is best to over-estimate the costs and be realistic with the sales predictions, both in terms of timing and amount. Do not, however, be so conservative as to reduce the revenue to the point of making the business become unattractive as an investment proposition, i.e. try to keep calculating the NPV and IRR of your company and make sure that you try to run the business in such a way as to keep the investors happy.

Revealing your confidential information

One of the pillars upon which your business will be based is in the confidential information that you generate, this being the intellectual property, marketing plans, etc. There is an understandable reluctance to reveal these documents to a series of investors who may be on the board of competitor companies by virtue of other investments. One solution is to send an extract of the business plan to potential investors in order to generate interest. This should contain all necessary information but not contain any information that you might consider to be confidential. However, the risk is that the investor may require more information in order to make a proper judgement and if so, you are bound to release it if you expect to get the money. Another solution is to ask potential investors to sign a confidentiality agreement. Prior to sending the plan to any potential investor you should do your own due diligence by investigating the nature of the investments that they have made to date and their policies with regard to the term and amounts of investment. By virtue of the reason you are approaching this set of investors, there is probably an interest in the business sector that you intend to enter and it is not unreasonable to suppose that they will have made investments in other similar businesses. There are various sources of this information, such as Companies House, the company literature (of both the investor and investee), business libraries (especially the City Business Library, London) and of course, your financial advisor should be of assistance.

Companies House in Cardiff holds records of those companies registered in England and Wales and there are satellite offices in Birmingham, Manchester and Leeds. Similar offices exist for Scotland and Northern Ireland. By post or by visitation you can view the following information (microfiche or photocopy) on any registered company; the accounts (remember to specify the year(s) required), the annual returns (the registered office, directors and company secretary details, information on shareholders), the memorandum and articles of association, any liquidation details and the register of mortgages and charges against the company. There is a modest charge for each of these documents.

What form of finance should you ask for and how do you treat inflation?

There are a number of financial structures that could be appropriate to your project and without a detailed knowledge of finance and the internal policies of the potential investor, it will be difficult for you to suggest a package. Thus let the investor suggest the form of the finance, perhaps even leaving both the amount and extent to a later stage. Remember, you do not have to accept it if the structure of the financing package is not suitable for your needs and you will no doubt have contacted other financiers anyway, so you will be aware of the types of deal on offer.

With regard to inflation, some include an allowance and some do not. For the sake of clarity it is perhaps better to discuss your finances in terms of constant prices, since inflation rates are difficult to predict accurately and could distort the figures, especially if different rates are chosen for different years. The absence of an inflation allowance may distort the calculation of net present value and internal rate of return, but this could easily be adjusted in a parallel set of calculations.

Preparing the business plan

The business plan must convey two main messages, which relate to: (a) the market opportunity; and (b) the ability of the proposed management to exploit it. Thus the first objective is to be realistic. Your target audience will have seen many plans such as your own and will have had some successes and some failures. They know the parameters. If you are unrealistic in any part of the plan, your credibility will be damaged. Furthermore, you must attempt to identify any strengths and weaknesses in your company and in particular, propose a strategy for minimising the effects of the weaknesses. If for example, you perceive a lack of experience in marketing, you could say that you will enlist the help of a DTI consultant, for example. Second, you must concentrate upon the skills of the management team and in particular the track record: how have they performed before? It is here that the plan conceived by the academic may be weak and the purpose of the following sections is to help you realise what other functions could be necessary for your business to succeed. You may be eminent in your scientific discipline, indeed you would be expected to be, but you may not have had the opportunity to develop in other areas. The maturity to identify and recruit in the areas of finance, marketing, etc. will greatly add to the credibility of your proposal. You should also include a section on how you cope with change, challenges and adversity. The plan should be professionally presented with a corporate image, lack of spelling mistakes, good grammar and an attractive but not ostentatious layout.

The plan must exude the characteristics that make both yourself and your business a special and return laden enterprise. In any event, your plan will be judged relative to others and at that time only. So, in addition to intrinsic factors such as your own ability, market potential in a given sector and the strength of your team, you must also address the opportunity investment issues, i.e. why should the investor put money into your sector, rather than any other and can you both create and maintain a competitive edge? All businesses, especially new ones, are faced with risks and as has been pointed out before, the degree of risk is related to the extent of the reward. Investors have a choice of risks in their investments and will judge your proposal accordingly. It is wise, however, to identify as many of these risks as you possibly can and to propose 'risk minimisation' strategies so that the investor can see that you have considered how things might go wrong. This section must not be selective, i.e. do not just include the risks that you feel you can counter. If there are those you cannot, either the investor may be able to help or you employ someone experienced in that area. Another technique that may be of use in containing risk is a sensitivity analysis whereby you provide a financial appraisal of the costs of the risk. Make sure the assumptions used have real-life validity and do not rely too heavily on your personal computer, whereby you could easily generate reams of relatively useless figures. A special form of sensitivity analysis is the break even point whereby you calculate the minimum amount of sales (or units to be sold) to cover all of your fixed costs. The breakeven point is useful as a management decision model but there are several assumptions that must be made. First, the patterns of costs and revenues should be linear over the range of output and that the costs may be split into fixed and variable elements. Second, that variable costs vary in proportion to the volume of output, the fixed costs remain constant and that the selling price does not change. The easiest method to apply uses the equation:

$$\text{sales} = \text{fixed costs} + \text{variable costs} + \text{net profit}.$$

At the breakeven point the net profit will be zero. If the price of a product is £200 per unit, the variable costs are £50 per unit and the fixed costs are £100 000 per annum, then the number of units required to break even (x) is given by:

$$200x = 50x + 100\,000.$$

Hence $x = 667$, i.e. 667 kits must be sold before a profit is made.

The overall presentation of the plan is crucial to determination of the impact it makes, but it is not necessary to make it over elaborate. Simple aids to the reader such as a contents page, executive summary, the use of figures, tables and appendices, and a brief, readable style will all be helpful. Begin with an executive summary of one A4 side, which concisely conveys the whole concept: a description of the business and markets, finance required, financial highlights to be expected, skills of the key management and milestones by which performance can be judged.

The contents of the business plan

The business plan must convey your business, your ideals and your passion for it. Thus the following sections are intended to give an indication of what should be included, but it is not definitive or in any particular order (Fig. 7.2). Many companies now derive a mission statement; a sentence which encapsulates their objectives. It is worth taking the time to carefully construct such a statement to preceed the executive summary, as it will be invaluable in capturing the readers interest in the pages that follow. The executive summary is perhaps the most important section since it is the first and perhaps only part of the business plan that investors may read, given the inordinate number of proposals that they receive. It

1. (a) Mission statement
   (b) The business opportunity
      Short term (1996–7)
      Medium term (1997–8)
      Long term (1998–2004)
2. Executive summary
3. Introduction to the business
4. Company history
5. Product strategy
6. Marketing strategy
      The market
      Intellectual property rights
      Marketing and sales
         Market positioning and competitor analysis
         Market research
         Pricing policy
         Geographical market segmentation
         Promotional strategy
7. Operations
      Project management system
      Quality control
      Premises
      Suppliers
8. Management and staff
9. Company structure and other commercial relationships
10. Company culture
11. Financial analysis
      Timing of funding requested
      Break even analysis/investment appraisal
12. Risks and rewards
13. Objectives and milestone

Schedule 1. Consolidated cash flow forecast, years 1 to 5
Schedule 2. Pro-forma balance sheet and profit and loss statement
Schedule 3. Detailed cash flow forecast, year 1
Schedule 4. First year budget

Appendices

Fig. 7.2. The contents of a business plan.

is inevitably the last section to be written and as such it is easy to spend too little time on it in the rush to finish the plan. This would be a mistake; you should endeavour to make the executive summary both powerful and persuasive with key financial indicators and well-constructed sentences. Your check list for the all important first pages is: (a) the business name, with an attractive logo if possible; (b) a mission statement; (c) the key contacts; (d) the scope of the business and its potential in the marketplace; (e) the resources required; (f) when financial returns are expected; and (g) historical data on the tack record of the management team.

Following the executive summary, there should be sections on the history of the company since its inception, detailed information on the management, the product portfolio, your marketing plans, a review of operations, including the project management system to be used, a financial analysis and a section on the risks and likely rewards, to include an appraisal of your objectives and milestones. Each of these will now be considered in turn.

☛ Company history

The rationale herein is to look at past performance as a guide to future performance. Since your biotechnology business is likely to be at an early stage, there may not be a great deal to mention, aside from the successful initiation of a research and development programme. There should be a summary of progress to date, with an indication of the amount of trading in contract research, services or products on the market. There should be a brief description of the key management together with the roles they fulfil in the company. Finally, there should be an indication of the current capital structure such as the source of equity, outstanding loans and the extent of involvement of shareholders and investors.

☛ The management

First, indicate the extent of control in the business exerted by the management, and the range of experience of the non-executive directors. Second, this should be followed by a stick and box type of organisational chart indicating the positions and key responsibilities of the management, together with any gaps to be filled. Then give a resumé of each person in no more than one-third to one-half of an A4 side, placing emphasis on the structure of the team. If a strengthening of the team is needed or an additional function is required, a similar short justification should be included. In an appendix, include full Curriculum Vitae (C.V.) for each person outlined above, but rather than the scientific version, try to tailor the text towards the following: (a) the aspects of the individuals achievements and experiences which are relevant to the future success of the company; (b) the relevant business experience; and (c) published papers, patents, awards and a personal history. At some point in this appendix, you should summarise the management with respect to salaries, equity stakes and job descriptions.

☛ The product portfolio

What do you intend to sell? In this section you should indicate the range of the companies products, such as contract research, services, kits, research tools, pharmaceutical products, etc. Very often this section will be incomplete and you will be raising money in order to develop new products. This is quite reasonable, although you should make sure that the reader understands the issues of: (a) tailoring the products to the markets you intend to address; and (b) being aware of the extent of the product applications. Also try to indicate why you think you have a competitive edge, i.e. identify your distinctive core competence. Describe the currently available technologies and products/services with which your product competes (if any) and compare their essential features. If there is no competition, try to say what led you to identify this market opportunity.

A separate section should now deal with intellectual property issues and the status of your patents, patents applied for or research that may lead to protectable material. The products should be classified as to stage, i.e. research, development, prototype, pre-production or production, and you should also indicate any clinical trials or regulatory approvals that might be required. This is particularly important if you are dealing with a potential new pharmaceutical compound.

☛ The product development cycle

At early stages in the life cycle of a product sales will be minimal for a variable period as the marketing plan is implemented, whereupon the cash generated for this product may be negative. In a multi-product company the effect of this is likely to be less than in a company with a narrow product portfolio. If you are in such a position you have two means to avoid bankruptcy: (a) raise finance; and (b) generate sales from a 'cash cow', i.e. a product that provides a regular income. The time scales in each of these areas will vary according to the market size, marketing budget, rate of technological change, etc. With regard to the latter, in the biotechnology sector the pace of change is rapid and you need not necessarily expect that the product launched to-day will still be viable in 18 months time. Similarly, it is not advisable to compare your product due for launch in 18 months time with those that are currently on the market. It is better to project your thoughts to products that might be available at around the time of your launch. Indeed, a discussion on the possible emergence of competitive technologies will add some weight to your arguments, as long as objectivity is retained.

Future products should over-lap with existing ones and are known as product line extensions. Besides the obvious new product development programme, you should be aiming to extend your current portfolio by developing existing products.

For example, adding new and improved plasmid vectors to a kit or providing an equivalent genetic testing service but for a new locus. In each case, you should be careful to identify the resources that will be required.

Market research and marketing your product

This should be the largest section of your business plan and is the one that requires the most creativity. Many do not spend sufficient time on marketing, to their inevitable cost. At first sight, marketing may seem obvious, trivial and a matter of common sense, but it is indeed a skilled and specialist area. To make a mistake in marketing is both noticeable and expensive, so the marketing strategy should be given a great deal of attention. In terms of cost, marketing budgets in the pharmaceutical sector represent an amazing ninety-fold increase over that spent on research. Obviously there will be considerable differences between marketing your biotechnology products and that of marketing a global pharmaceutical, but suffice to say, make sure you correctly balance the marketing effort against the stage of development of your business and do not underestimate the resource required for a comprehensive marketing campaign. In considering the elements of the marketing mix (price, product, promotion and place) it is important to remember that all four criteria are linked, the pricing level will be in part determined by the extent of promotion. The geographical place of sale by the product and so on. Defining a marketing strategy is thus an iterative process to be consolidated over a period of acquiring market information.

Pricing

For contract research projects and provision of services, it is highly recommended that you follow the designated man-year (dmy) policy outlined earlier. This is in reality a price minus policy, whereby you find the amount that it would otherwise cost someone else to do the project and charge slightly less than this. For large pharmaceutical companies this number is in the region of £150 000 per dmy and you should price your contracts accordingly. Under no circumstances should you be drawn into discussion of costs and allow your profit margins to be revealed. The latter is indicative of cost plus pricing whereby you add up all your costs and add on an acceptable profit margin. The danger is, of course, that you fail to identify all relevant costs thus eating into your profit margin and at worse, making the project unprofitable. The pricing of a commodity such as a research kit, enzyme or antibody should be according to industry norms. This means that you should identify similar products in your sector by, for example, examination of catalogues produced by other companies and price accordingly. Hopefully your competitive advantage, i.e. the improvement you have devised will enable you to offer a lower price per unit. However, if there are no similar products in the market, you should first saturate initial demand with a price skimming strategy, in conjunction with an extensive marketing campaign. This means that you charge a high price for the product, selling to those who are prepared to pay a premium to gain advantage. Such a strategy should be continued as long as possible, until competitors emerge and force you to charge a lower price.

☛ The product

The product should be carefully defined and may take the form of a service, a commodity, consultancy, contract research, etc. You should also try to identify what your next product will be, perhaps a derivative of the first (a line extension) or something new (a range extension). This is crucial for your strategic planning and will: (a) help you keep ahead of the market if competitors emerge; and (b) help convince investors of your desire to build a business that is based upon more than one product, where the risks of failure are high. Be careful to take a creative look at your product range in the broader market and try to ensure that all areas are covered. This means, for example, that you have considered the relevance of your product to all therapeutic areas, to the stage of the drug development process, etc.

☛ The place

Where do you intend to sell the product? The first consideration is the geographical location. Biotechnology products have a global appeal and you should perhaps first consider the UK/European market, followed by entry into the US and Japanese markets. Whether this is under your own initiative or in a strategic alliance/joint venture is secondary at this juncture.

If you wish to sell your product or service overseas then the DTI has an excellent 'export initiative' service. Begin by contacting your local business LINK office after obtaining the address from the DTI if necessary. The initiative is unlikely to become involved unless you can demonstrate credibility, e.g. you have a strong IP position. There are several routes to obtain help. (1) The New Product Service, whereby the company will be assessed as being fit for export, i.e. understands the export process and has a valid IP position. This is a highly subsidised service costing £60 for the first country and £30 thereafter. (2) A Market Information Enquiry, whereby a specialist will undertake a market analysis in the chosen country.

☛ Promotions

The ways in which you can promote your product are legion, being to some extent limited only by your own creativity. Some are obvious, others less so. Most require a considerable degree of hard work, dedication and credibility. In designing a promotional strategy you should adopt a portfolio approach whereby you expect several initiatives to each play a part in the success of the campaign. Having all of your hopes in a single arena is less likely to result in success.

You should first consider producing marketing materials, such as a brochure and price list. Much of the work can be done on your own personal computer, but it is as well to include some professional photographs since the brochure will probably be the first interface your company has with the customer. The front line approach should be a direct approach to companies within the sector of interest, using personal contacts (networking) wherever possible. Second, a mailshot should be prepared and will probably represent your major marketing effort, but it must be precisely targetted. It should consist of a concise, direct and informative letter of introduction, a non-confidential summary or brochure and perhaps a price list. However, before undertaking this campaign you must be sure that the IP position is as secure as is possible because greater awareness of your company in the outside world will also bring any detractors to the fore. In preparing the mailshot it is worth investing some time and effort into the construction of a customer/contact database using a suitable computer package. This is an invaluable asset to the company and will save much time and effort if set up properly.

A good method for selling your product or service is to use an agent, a person or company who is contractually obliged to sell these wares to third parties. Often you can select a specialist in a particular market or geographical segment and rely upon their network of contacts for new business. Around 5% of the contract price would be a fair fee, but of course this should be added onto the selling price, so there should be a minimal impact upon your profit margins. It is essential to have a lawyer draw up an agency contract for you, since the laws relating to agency agreements are complex and require explanation. Agency agreements are a good way of reaching distant markets for which you may not have the resources to conduct a useful marketing campaign yourself. It is almost essential to use this route if you wish to impact upon the Japanese market. In particular, recent changes now require there to be: (a) miminum notice periods; (b) payments to agents upon termination; and (c) regulation of the entitlement to commission. In addition, there are also implied duties for both the principal and agent.

Another useful marketing tool would be to hold an academic symposium around the subject of interest, making sure that you invited prospective customers. For example, you could invite key speakers in the field to give a lecture and accompany this with a lunch sponsored by your company and commercial presentation. Grants from the local Training and Enterprise Council may be available for such purposes. A survey of relevant conferences should also be made, with a view to presenting a stand that promotes the relevant product/ service. These facilities are usually quite expensive so you may have to wait until you are succesfully trading, but if targetted correctly, i.e. to a conference with a significant commercial attendance, then the results could be pleasing.

Direct advertisment in trade journals (e.g. *Nature*, *Biotechnology*) is a highly effective method of promotion, especially if you can obtain, for example, 'New

Product Focus' status. In addition, there are several public databases which could be used (e.g. Discovery), that are screened by the pharmaceutical industry.

☛ A description of the market and its prospects

In this section of the business plan you should introduce the general markets into which your company intends to operate, by providing a general description, an indication of its size, the relative growth rates both during the past five years and those forecast for the next five years, together with the potential customers, whether these be institutions, consumers, government organisations or manufacturers. If you have performed a market analysis survey, it should be presented at this point. If the product or service represents an improvement on existing technologies then quantitative statements of same should be presented.

Markets are not uniform with respect to opportunity and any given product or service is likely to find a number outlets in different market segments. The next task, therefore, is to precisely define the market you intend to penetrate, i.e. carefully segment the market. Indicate their individual current sizes, relative projected growth rates and geographical characteristics, whether these be regional, national or international. Take care not to use statistics that relate to a wide market (e.g. those which may be derived from newspaper data) for extrapolation into the narrower market you intend to address. Also, try to establish any unusual market characteristics and clearly state how these may impact upon your business. For example, is the market dominated by one or a few large players such that penetration will be made more difficult or will your chances be enhanced by entry into a fragmented market but your profits reduced due to the intense competition? Try not to assume that each market segment that you identify has an equivalent mixture of small, medium and large companies, since this will profoundly affect the percentage of market share that you can expect to gain. Failure to note the latter point can lead to erroneous sales predictions. If you have a fixed overhead structure, as is often the case with campus based start ups, then there will probably be insufficient flexibility to cope with over optimistic sales figures, i.e. your business will be seen as a failure, on the basis of a lack of attention to market analysis.

For each market segment a separate plan of approach should be devised, being careful to define: (a) how customers in each segment take the purchasing decision (e.g. unit purchases, long-term contracts, competitive tender); (b) the nature of the order (size, approved suppliers, sourcing methods); (c) at what level in the organisation the purchasing decision is made; (d) what are the critical characteristics of the product for the segment in question (e.g. quality standards, performance measures, availability, servicing and pricing); and (e) are there any special circumstances such as seasonal or cyclical purchasing patterns?

☛ Identification of the competition

The identity of the competition to your business will have a profound effect upon your chances of success and it is essential to think very hard about who they are, their strategies and likely response to new products entering the market. It is not valid to assume that your product is so superior that there will be no competition, especially if you are a small biotechnology start up entering a market where there are companies with far greater technological and marketing resources than your own.

The most common mistakes are to over estimate your own competitive strengths when writing your business plan and to underestimate your weaknesses. Investors will be looking for realism in your analysis of the competition. The second error is to not fully consider the impact of competitive responses to your product in the market place. If your new product or service directly threatens the profits or market share of a major competitor, then you can expect an aggressive response. The effect of this on your own profits can be considerable, which will be reflected in a failure to achieve the profits expressed in your business plan. Therefore, build such a contingency into your three-to-five-year cash flow projections. The analysis should not just include existing competitors, since new rivals will appear over the first few years of trading, especially if your product or service is a good one. These comments are particularly applicable to the immature and dynamic markets that can be found in the biotechnology sector. Whilst it is impossible to predict how many new rivals there will be and the extent of their economic impact, a contingency should be included in the business plan.

Another key factor is the extent of the market share that you think will be attainable. As businesses and markets mature, so the fight for market share intensifies and although it is of lower importance to a new biotechnology company in an immature market, your investors will be expecting to see it included in your analysis. Thus make an estimate of the market share you expect to attain in the three- to five-year timespan, taking account of existing and likely new competitors and including a rationale. Define the niche you expect to fill and indicate from which competitors you expect to draw market share and why. This will also give an indication of the likely source of competitive response. No doubt you will mount a counter response to any attack on your newly achieved market share and thus at some point you should analyse the potential strengths and weaknesses of the competitors, directly comparing your products and services with theirs.

☛ Sales and marketing

The marketing strategy is the most important part of the formula for success. Unfortunately it is often underestimated in importance by entrepreneurs and investors alike. There is no doubt that you should always make your business market driven, no matter how technologically brilliant you believe the opportunity to be. Some companies raise venture capital on the basis of a research and development programme that may take five years to come to fruition. With continued financial support there is time to devise a marketing plan but this will undoubtedly require the addition of new personnel to your team. If you believe your new venture to be more downstream, i.e. you may have products near to market, then you must indicate how your potential products and/or services will be positioned relative to those of competitors, with particular regard to pricing, the quality of the product/service and the image you wish to present. The pricing strategy that you adopt is fundamental to your success or failure and you are strongly recommended to seek advice from an expert in the field. The majority of those from an academic background will see pricing as an extension of a research grant, being the total of the costs (direct and indirect) plus an addition for profit. This is absolutely inadequate and will probably result in you severely compromising your cash flow and profitability. Some of the more obvious reasons are: (a) that you may fail to account for all of the costs; and (b) that you underestimate the market value of the product and are unable to satisfy demand. For general purposes in an immature market, the most workable strategy is that of the price minus approach whereby you offer a lower price (e.g. 10–15% lower) than that of potential competitors.

In your analysis you should allow for the response of your competitors: will they reduce price, offer a better service or increase the price and try to capture the quality end of the market? For example, there is one particular manufacturer who charges a high price for restriction enzymes. However, the image created is one where the enzymes will always function to expectation, are stable and are nuclease free. What price a failed experiment due to inadequate enzymes?

Whatever your market positioning, a comprehensive marketing plan must be developed, which should be presented in only a shortened form in the business plan. Thus you should prepare a separate marketing document that the investor can peruse at a later point, including realistic, quantitative market research data and market analyses.

The sales strategy is also a vital aspect of the business plan and in many small biotechnology start ups it is an area to which too little attention is paid. Even if your initial plans are for a five-year R & D programme, you should still attempt to define the sales strategy. The projections for sales should mirror the targets derived from the market research data. The first decision concerns the nature of the distribution channels you intend to use. For the most part these will be a mixture of your company's own direct sales efforts, the use of expert services or use of agents with designated territories. If the market is fairly narrow and well defined, then you may be able to immediately enter an international market. For example, if you were selling a service to the pharmaceutical industry then the global nature of many of the companies therein would necessitate a global approach to the marketing campaign. In this case you should indicate exactly how you intend to reach these customers (e.g. by personal visits/seminars, use of the

DTI export scheme) and indicate an adequate budget for this activity. Do not underestimate the costs of a marketing campaign. It will also be useful for the reader to understand the frequency of contacts you intend to make and how you intend to follow them up. A rough guide to the size of order you might expect will also be useful, although this is again very difficult to predict.

It is not sufficient to project sales by assuming that you will gain a particular market share of an assumed market size. This must be accompanied by a precise definition of the market and a strategy for how you intend to sell it there, plus an estimate of the probability of success. As before, contingency and 'what if' calculations should be included showing the effects of possible breakdown points.

☛ Operations

This section of the business plan relates to how you will produce the product or manage the service that your company is offering. If you were involved in a manufacturing company this section would be quite extensive but for a start-up biotechnology company it can be brief. Depending upon the anticipated position in five years' time, when perhaps you may have a drug in clinical trials, then an indication of how you would manage this process and the finances required would be helpful. However, these issues are likely to play a more significant role when considering the expansion of your business utilising second round financing.

For early stage businesses you need to indicate the project management system you intend to apply and how you will monitor project progress. This should be either in the form of a GANTT chart or a critical path analysis, with a full appraisal of the resources required at each stage and the timing and interdependence of these stages. Project planning is a complex task and if done well it will take a considerable period of time. The GANTT chart is a graphical representation of the projected progress of a project over time. It is usually presented in a horizontal bar chart format with annotations to indicate the resources required at each point. A critical path analysis is applicable with more complex projects that require the interaction of several pathways, e.g. when part A must be completed before parts B and C. This method of project planning is often presented with a 'stick and box' format and it is useful for identification of the steps that are limiting to the timely completion of the project.

The operations section of the business plan should also indicate the extent of your dependence upon key staff (e.g. skilled technical staff), on the acquisition of key materials (e.g. licensing arrangements and terms), on subcontractors and on physical parameters such as equipment and premises. For example, small premises may limit your ability to expand if your business plans are successful. How would you cope with a quantum leap in capacity that resulted from new premises?

☛ Finance

The finance section should be used to integrate all of the predictions made in the preceding paragraphs, preferably in summary form and concentrating on the overall structure of the company. For example, 'R & D spending will lead to losses until year 3 when sales of services will begin to increase, leading to profitability in year 4'. A full financial analysis should appear as an appendix covering a five-year period, this to include sales figures, profits before tax, the impact of your marketing strategy, the R & D budgets, a predicted (pro-forma) balance sheet, profit and loss statement and cash flow projections. The latter is of paramount importance since without the expectation of cash, the business will not survive. The balance sheet is a statement of the source of the funds, which is expressed in terms of equity participation, loans, retained profit and how these categories are allocated. This will include fixed assets such as premises and capital equipment and current assets such as working capital, work in progress, stock, etc. There should also be a statement of current liabilities, i.e. monies owed to creditors, tax liabilities and bank overdrafts. The profit and loss statement tabulates the gross sales income from which all relevant costs are deducted. First, the cost of goods sold is deducted (to include variable costs such as materials and labour) to give the gross profit. This does not include the cost of premises or capital equipment. Second, deduct operating expenses such as rent, rates, depreciation, marketing costs and administration to give the profit before tax. Third, deduct tax, interest charges and directors fees in order to leave the profit after tax, which can be distributed as required. As with the break-even calculation, which should also be included in this section of the business plan, you should perform a sensitivity analysis. For example, test the financial forecasts with a 20% reduction in sales income and a 20% increase in costs. Will your business still survive?

This information should be followed by an analysis of the funds required for the financing of your business, to include amounts, the timing of when the funds are required and how the money will be used. It is less than advisable to indicate what type of money you require, which is a decision best left to the financier you are approaching. Since the exit route will be high on the investors agenda, you should also perhaps indicate how you think they may realise their investment, e.g. by entering the stock market (public offering), sale of the business to a third party, etc.

☛ Rewards and attendant risks

Summarise the relationship between the risks and the rewards that have been discussed throughout your business plan. Indicate the major risks to the business and whether they are within your influence. If the latter, state how you will attempt to minimise these risks, e.g. by use of a particular managerial style. Follow this up by an indication of the value of the company should you achieve your business plan. If the business is based on a particular piece of intellectual property then this will give the business an opening value that should be built upon as the business develops. For example, erring on the prudent side, a patent to a generic technology should be worth £0.5m. Within three to five years, you could expect the value of the company to be £2.5m. The five-fold increase in company worth will be reflected in share value and when stated as such, may show an attractive investment.

☛ Summary of objectives

Using the GANTT chart or critical path analysis, it will also be helpful to summarise the milestones and objectives for the business, giving an integrated appraisal of the timings and deadlines for each segment of the business.

☛ Appendices

An appendix should be used as a vehicle for allowing large amounts of information not to be included in the body of the business plan, whilst still giving the investor access to relevant information should it be required. Included may be: evidence of intellectual property, market research information and a detailed marketing plan, details of current shareholders, an organisational chart with an indication of how it may develop, curriculum vitae of key staff and financial projections for three to five years hence. The latter could include a proforma balance sheet, profit and loss account and cash flow projection. If you have audited accounts, then these should also be included.

Business plans for venture capital

In general, venture firms will not begin to appraise a project until they have a business plan. It is probably your first line of approach and given the number of proposals reaching the venture companies, it has to make an immediate impact. In an executive summary (one A4 side) you should convey the magic of the idea, the commercial potential (accurately analysed), a meticulous plan of operation and the quality of the management and teams required for its execution. Clearly, this is not an easy task.

Your business plan may be the first and only financing proposal considered by the potential investor on your behalf and as such, it has to be correct. Only around 5% of proposals reach the negotiation/acceptance stage, the rest are rejected. Of these, only 10% are destined to succeed commercially and as such, we can see that the investor will loose the investment on the 90% that do not. Thus in a risk based market, returns must be high to be attractive. It is not advisable to use one investor to 'review' your proposal in the hope that you can obtain useful comment for the next investor; the community is small and comments as to your personal quality and credibility may soon circulate.

The business plan for technology based businesses

Investors will be faced with proposals from many sectors, not just your own. Whilst your proposal may be technically excellent and would survive the harshest scientific criticism, the issues do not just involve technology and there are additional areas that the potential investor will concentrate upon. If presenting to a bank, science associated proposals will only be a very small proportion of those that are seen. How then to bridge this gap? First, you must gain the interest and enthusiasm of the potential investor, by including at the beginning, the items they expect to see, for example, a summary with the amount of the loan, nature of the product, marketing strategies and cash flow predictions. Technical jargon is likely to cause a rapid loss of interest, no matter how good it is. Thus in approaching banks and other such sources of finance, it is essential to remember that the first people you are likely to meet are non-technical and will not understand the science associated with the company. It is thus better to focus upon the implications of the technology, not what it actually is. Better still, orient these implications towards business potential.

Simplicity is the key to the presentation of a business plan and it is quite an exercise to communicate your science at a basic level, when one is used to scientific literature and departmental seminars. For the venture community you can afford to be a little more complex but again, the first line of presentation should be non-technical. Smaller non-specialist seed funds should be treated like banks, larger funds will have technical specialists waiting in the wings. Thus, in the latter case it may be better to prepare the business plan with a brief technical description and subsequently add more detailed relevant scientific information in an appendix. The latter should be as hard hitting as possible, because the scientific analysts employed by the top venture funds are extremely good indeed. This approach will have the advantage of not making the business plan overly top heavy with technical information.

A second problem, which is often apparent in technological-based business propositions is the shortage of a complete range of management skills. This may not be apparent to you in writing the proposal, but you will need a significant number of other skills in order to get a business off the ground. It is perhaps symptomatic of the scientific community that they are often blinkered to other issues and reject the idea that they do not have these skills or see them as some kind of trivial 'add on' that can be summoned at will. This often reflects basic insecurities and a lack of training in team-based activities, which will be explored in a subsequent monograph. However, these deficiencies will be all too apparent to the investor and you will have to be very convincing on the issues related to management of the finance, management of the operations (how you will do it) and in particular, the marketing strategy. The business proposals that you have prepared will need to have strong representation in the latter areas. Since investors are likely to perceive this as a weakness, simply because you are a scientist, then your strategy and numbers need to be significantly better than the competition.

A third problem is in marketing. The most difficult area to enter in any sphere is 'new product, new market'. In biotechnology based businesses, the problem is compounded by the rapid change in the market place. This is especially true with reagents and kits for diagnostics and/or *in vitro* use and whilst longer-term projects may seem more attractive, the rapid technological advances leads to a significant risk of obsolescence. Investors are aware that the window of opportunity may indeed be short, which means that your business plan must be carefully prepared and presented. It will require about 200 hours of work!

8 Financing the business start up

We have no money. Therefore, we must think.
Ernest Rutherford
From a speech made circa 1920

Introduction

The licensing option may or may not be suitable for your needs. Perhaps your aspirations and view of the market opportunity now dictate that you should 'go it alone' and consider starting your own company. It may be that you are in a position to set up a company and licence your technology through it, rather than your host institution. Whatever the circumstance, you will find that your business wants one thing more than any other – cash! The nature and type of financial arrangements that are available to you will be dictated by the purpose for which you require the money and the degree of risk involved for the investor. Investors are not in general of a charitable nature and will require a return on the investment. The more risk involved, the more justification and analyses you will be expected to provide. There are several mechanisms for financing the venture which are all dependent upon the degree of risk.

☛ Short-term or long-term finance options

The time period of the investment is of significance, since all other factors being equal, shorter investments are likely to have a more accurate analysis associated with them. The variability of market conditions makes sales and cost predictions inaccurate beyond two years and thus an assessment of market risk will be necessary.

☛ The use of security to reduce risk

The investor will often require security on the investment. Often referred to as personal guarantees (PGs), this may involve any of your tangible assets, such as your cash, house etc. In the loan agreement such goods may be forfeit if the business fails. From the investors point of view, the logic is clear: why should they be prepared to invest their money if you are not prepared to invest your own? Some may even conclude that the extent of the security you have to offer is indicative of your commitment to the business. In this regard, a security of £10 000 is likely to be more politically convincing if it is all you have, compared to a similar investment from someone who has £100 000 or more. It is known to investors that such a commitment can lead to both a stressed management and inappropriate managerial decisions that result from short-term financial pressures and as such, they are becoming less popular. However, it will still be expected that you be prepared to put up some form of security, even if in the final analysis the investor can be convinced of your commitment and decides not to ask for your PGs.

☛ Sharing Equity

Financial institutions may be prepared to take a stake in your business, and share the risk, by making an investment in your company in exchange for shares. Presumably you, your colleagues and/or your institution will own 100% of the company, i.e. 100% of the **equity**. The extent to which you participate in this route is dependent upon the extent to which you are prepared to give up some of the **control** of the business. There are several points to consider. (1) That 100% of zero is still zero. Without the investment you may be unlikely to make any money at all, but you will still retain all of the control. Our previous discussion of cash requirements indicates that the majority of new businesses (or indeed new products within an existing business) run cash negative for a while, in order to support research and development, sales and marketing costs, etc. It may be that this cash injection enables you to overcome the activation barrier and get your product to market, i.e. a reduction in your share capital may result in an increased financial return. (2) Depending upon the size of the investment, the investor may require a degree of management control, e.g. a seat on the board. This is a point for negotiation, but you should consider the consequences. How much control and influence will this company exert? Experienced business managers could be a valuable asset to you, but may wish to change the direction of your business. Is this appropriate? How prepared are you to defend your market research and marketing plan? Does the investor have a short- or long-term view? What happens if the personnel changes? Many companies will retain the right to appoint whoever they see fit or indeed, the person with whom you have negotiated may leave his/her employ. You could then find yourself dealing with some very different characters. How experienced is this person in the areas that you and the business requires? Do they bring financial, marketing or any other skills?

Use of working capital

A lack of cash to service your immediate debts could lead to bankruptcy. However, the amount of working capital both available and necessary varies considerably according to the state and nature of the business. Working capital may be required to invest in some new staff or expand a research area (more consumables, legal costs, etc.). In these cases it should be relatively straight forward to justify the investment. For example, a new contract that has gone to Heads of Agreement may require you to take a risk and prepare for the work to be carried out, to reduce lead times and make sure you have the best chance of that all important success fee.

Your aspirations and business objectives

The financing route may also be influenced by the pace at which you wish to develop the business, whether this be a requirement for finance to: (a) maintain the *status quo*, perhaps in adverse conditions; (b) undergo a 'risk minimisation', 'steady but sure' growth programme; or (c) go for the 'big hit', i.e. rapid growth, high returns, high risk. For the most part, commercialisation of biotechnology will involve the latter, taking account of the untested and unproven markets and rapid scientific progress, with the concurrent risk of being superceded.

Sources of finance for your business

The four broad categories for raising finance are: (a) sales of products or services (e.g. contract research or royalty income); (b) debt finance (for example, from a bank); (c) an equity injection (e.g. from a venture fund); or (d) other sources (e.g. a government or EEC grant, loans from family members).

It is essential that your funding requirements have been meticulously planned; this is probably the most difficult area to get to grips with for those from a scientific background. It is strongly advised that you get sound financial advice from an accountant with some experience in these matters. This text is intended to give you sufficient knowledge for a sensible conversation with such specialists. It is not sufficient to look at your projected (*pro forma*) cash flow and assume that the maximum amount required is the largest negative figure and then propose a rather woolly split between a bank loan and an issue of share capital. There are four questions that must be specifically answered. (1) Why is the money required? (2) When is it needed? (3) What type of money suits your purposes best? (4) Why should the investors invest and what are the routes by which the investors can realise their investment (exit routes)? The latter issue will be paramount to the investor and you must give it considerable attention. Before discussing financing options in detail, the above questions will be addressed.

☛ Why is the money required?

This must be clearly stated since although *you* may have a good idea as to why the money is required, the investor may not. Furthermore, it may not be as apparent from your business plan as you might think. A clear presentation will help the investor see what types of money are needed. In writing the proposal you should be very clear as to the purpose of the financing requirement. Examples are as follows:

1 Funds for developing a product after the initial research has been done. In this case, you will have identified a potential for a product as a spin off resulting from your normal research activities. However, you may need someone to develop it further, e.g. prepare quality control sheets, sort out packaging, etc. In addition, you may require funds for commercialising the product, e.g. legal fees for drafting contracts, product literature, advertising, etc.
2 If the business is already moving, you may wish to develop a new generation of products (e.g. improved plasmid vectors with more convenient restriction sites) and enter new markets. Again you could require funds for researching into these markets, finding suitable clients and sorting out who the competition is.
3 A loan may also be used to provide you with working capital to finance day to day activities. For example, you may have a lot of trade and pay your bills on time, but there may be those who owe you money and have been slow in paying. This could generate a problem in the cash flow of an otherwise healthy enterprise. A loan to tide you over the rough period (providing you are chasing your debtors) is a perfectly laudable reason.

☛ When is the money needed?

Whilst your budget will indicate the amount required over the first year, this may not be required all at once. This means that you do not have to assume the entire debt plus interest at the outset and that the investor does not risk all of the money at once. Your carefully prepared milestones will be essential for measurement of performance as the venture proceeds, so that if variations to the funding structure are required then this can be accommodated without surprises. Some of the capital equipment and ready cash will be required prior to trading, i.e. to prepare the product for market. The initial requirement could be further reduced by hire purchase of capital equipment, for example.

☛ What type of money suits your purposes best?

To a great extent the nature of the finance that is appropriate will depend on the use of the funds. Various mechanisms have evolved to suit particular purposes and it is advisable to stick to the tried and trusted routes. Purchase of fixed assets (e.g.

centrifuges, autoclaves, etc.) and an increase in working capital (consumables, legal fees) is a relatively secure investment and can be dealt with in a straight forward manner. If you have security to offer (e.g. remortgaging your house) then this opens up the realm of debt financing via the bank. Different funding mechanisms are possible for short- and long-term requirements and will depend upon: (a) the security that you can offer; and (b) whether you are prepared to give up equity in the business to someone who is prepared to take a risk on your venture. In addition, there are different types of funding according to the degree of risk. The safer options of easing you over a working capital requirement or slow steady growth are not likely to interest the venture capitalist. They expect greater returns from what would be considered greater risk. This will come from growth and expansion of the business, but of course will inevitably result in creaking at the seams.

Short-term finance

☛ Working capital

Working capital is the daily cash requirement required for general running of the business, e.g. paying for materials, salaries, etc. This can be obtained via a bank account with a clearing bank and a negotiated overdraft facility if necessary or via discount houses or factoring companies. The high street clearing banks are major sources of finance for small business and manage short-term borrowings (overdraft) to accommodate cash requirements. Besides the big four banks (NatWest, Barclays, Midland and Lloyds) there are many others that could help with sources of small business finance. They offer other services as well, including access to the government loan guarantee scheme, a service to help with export and import of goods and for the riskier projects, a venture capital facility. The size of the overdraft must be pre-arranged and never exceeded, since bankers do not like surprises. The extent of the overdraft will depend upon your current cash flow position, business risk and the amounts tied up with creditors and debtors. Obviously the bank will be charging interest on the overdraft, from which it makes a profit. However, a bank will not allow the facility if the business is excessively risky or poorly run, since there is the risk of losing the money entirely. Hence the well-written business plan becomes a valuable working document.

Factoring is a service which provides funds based upon outstanding invoices. The factoring company buys the invoices in exchange for up to 80% of their value and pays the rest upon collection, minus a service charge. As with an overdraft facility, the extent of factoring cash may increase with the trading position of your company. The companies suitable for factoring are usually small, with a minimum turnover of £250000.

Discount markets work with a Bill of Exchange, similar to a post-dated personal cheque. Bills can be drawn up and sent with the goods and the receiver must pay by a given date. The Bill can be sold to either a discount market company or a bank in return for cash. Payment is in relation to the credit status of the customer (thus reflecting risk) and a discount representing the fees. The latter are quite competitive with the interest rates that could be charged on a bank overdraft and using such a financial method will release the bank overdraft for use in other arenas.

☛ Bank loans

Depending upon the credit rating, interest charges will be between 1.5 and 5% above the base lending rate, being the other principle way in which the clearing banks generate revenue. As a new business start up, you will be a credit risk so this can be expected to be on the high side. However, the interest rate is negotiable and you can always approach different banks to get some quotes.

☛ Hire purchase

Under a hire purchase agreement, capital equipment can be obtained and paid for over a pre-agreed period, with the assumption that you will pay a final nominal sum and the item will be yours. The arrangements usually involve a 10 to 30% deposit followed by monthly payments. Again this reduces risk for the business since: (a) you now have assets that have a realisable value should you go under; (b) the investor need not expand all of the money at the beginning; and (c) if you lack acceptable security, the capital equipment will provide it. This could be of particular importance to the academic who is starting a biotechnology concern. There are several finance houses that you may find amenable so it will be wise to explore several to get the best deal.

Longer-term funding

☛ Long-term loans

Loans can be obtained from clearing banks for periods of up to 20 years with various options as to the interest rate. Security will be required, either against premises or capital equipment or against personal guarantees, such as your house. If you have invested £30 000 or more then it is a little unreasonable for the bank to ask for your house, no matter what the size of the loan. The interest rates are negotiable and again, it is worth approaching different banks to secure the best arrangement. In addition, it is currently common practice for the bank to grant a capital repayment holiday, i.e. a period in which you only repay the interest. This could have a major impact on the cash flow and be very useful during the early stages. Larger amounts can be borrowed from specialist merchant banks and such an association could be very helpful if you wished to 'go public'.

☛ Debentures

For a limited company, debentures can be used as a long-term loan upon which interest must be paid. Debentures can be issued like shares and will be secured against a particular piece of equipment, which may have to be sold if interest charges accrue.

☛ Shares

Limited companies can also issue shares to generate cash. A successful biotechnology company might enter the public arena via an **initial public offering** (IPO) after several years trading. The rules for entering the stock market are particularly strict in Europe, although the US may be an easier option. Shares can also be issued privately to individuals or to venture capital organisations. The latter can raise finance at all stages of business development, right from seed finance (less than £100 000) to larger development funds of several million pounds. These investors will be looking for an exit route and it is up to you to facilitate this pathway. Beware that the issuance of shares may well dilute your control of the business.

☛ Commercial mortgage

Should your biotechnology company be successful and you are fortunate enough to own the premises, then these can be mortgaged to release cash.

Why should an investment be made?

Investors will put money into your company if they think it will make money for them. Otherwise, why do it? The banks route for realisation of the investment should now be clear, interest on the overdraft or term loan. They seek to reduce risk by asking for loan security but otherwise will probably not foreclose unless you do something irresponsible, like repeatedly exceed your overdraft limit. For equity investors (e.g. venture capital) the questions are more complex, partially because large capital gains will be expected on the three to seven year time horizon. The two key issues are: (a) the amount of equity that you feel prepared to 'give' away; and (b) the exit routes. It is unlikely that there will be a requirement for total control of your business operations but some venture capitalists will want a seat on your board, seeking to influence business decisions by that route. This is largely because they will be investing in a number of companies and will have neither the time nor resources to run your venture. However, a condition of the equity participation is likely to tie you to the business plan as originally presented. Significant departures at a later juncture (e.g. if things are not going well or other markets requiring new R & D look interesting) will probably not be acceptable without prior approval.

Emotionally, many people will find it difficult to give up part of their business to an outsider. Such feelings should be suppressed, since you require the investment to develop. Alternatively, from the investors point of view, why should money be given for nothing, especially to a possibly unproven management team, risky technology, untested market, etc.? Herein lies the reason for requiring a large **return on the investment**: it is to compensate for the failures!

The investors will be driven by investment criteria set by their own companies and from the beginning will have identified suitable exit routes, i.e. ways in which they can realise their investment and free funds for other ventures. This may differ from your own goals that may be both financial and non-financial (a long-term stable income for your family, employment freedom, the satisfaction derived from running your own business, etc.). If you accept the money in the first place you should be aware that the investors will eventually wish to cash in. You can prepare for this by, for example, arranging for further development finance or making plans for a new business, etc. There are primarily six routes by which the investor may cash in or cash out, depending upon your point of view: (a) selling the company to another company; (b) going public, i.e. listing on the stock exchange; (c) selling the shares back to the company; (d) reorganisation and rationalisation, e.g. selling off parts; (e) selling the shares to a new investor; or (f) liquidation. Which route is taken will largely depend upon what is appropriate at the time.

Cashing in or cashing out: exit routes for the investor

The nature of the probable exit route will have a profound impact upon both yourself and your business, so it is good to appreciate the relevant parameters.

☛ Selling the company to another company

If the whole company can be sold then you are likely to get more for the business than an equivalent portion of a minority interest in the company. The issues lie in control of the business. If you still own 51%, it is unlikely that the board will be able to force you to sell all of the company. However, if a venture capital fund (or a consortium in agreement) owns more than this, sale of the business in its entirety is possible. The simplest method is to sell the business for cash. In some instances a special form of financial instrument, the note, may be offered. This is undertaking to pay cash over a period of time. This method should be viewed with caution since it is entirely possible that a buyer could ruin your business and then not have sufficient cash to fulfil the obligations of the notes, i.e. you should make sure the notes are secured. This method has tax advantages. When being bought by a larger company, there is the option of taking stock in the larger company. Since most 'stock exchange' arrangements are tax free, this could be a good option. As a note of caution, it may be wise to take the stock in a public limited company rather than a similar, privately held company since you will not necessarily have income nor equity from the latter. Also, it would be wise to obtain registered shares such that stock can be sold any time you wish, preferably dividend paying preferred stock. These can be convertible into common stock for sale whilst still giving you income. This method has numerous consequences for you and your business but, if handled correctly, is an effective exit route and one which is likely to be acceptable to your investors.

☛ Going public

The 'public' will pay more for your business than anyone else. It is the ideal and preferred exit route and could leave you in control. However, it will only work if your business has been successful and it should be the long-term goal when you write the business plan. It has the advantages of: (a) you being able to sell some of your shares and not necessarily all of them (which applies to investors too); and (b) the possibility of raising more cash at a subsequent point by a further issuance of shares. By the time you are considering 'going public' you will have raised money from various sources, probably including merchant banks. Both these and the larger venture capital companies have specialists in organising the offering and will be able to help. Indeed, they will be seeking to maximise their return and minimise errors so it may be an idea to take a back seat, observe closely and be carried along in the wash. In order to go public you will have to demonstrate an excellent earnings track record and all criteria for good projected earnings. Exactly how many of your shares can be sold at an initial public offering (IPO) will be decided by an underwriter (stockbroker) who will charge a fee. Higher risk businesses will result in higher fees although this may be reduced by negotiating the stockbroker some share options. In general, the fee will be between 5 and 10%.

☛ Selling shares back to the company

Investors may indicate to you that they wish to sell their shares in your company back to you. This will give additional control, may remove a 'difficult' investor from your hair and could free up some shares for sale to another investor. For example, an investor who has taken a risk at an early stage may wish to realise part of the initial investment leaving you free to attract a larger player. How do you finance this? The two options are cash in hand or a bank loan. The latter will have the result of converting equity into debt for a short period, thus increasing the liability of your company. The third exit route in this category is by 'puts and calls'. A 'put' will be negotiated in the original investment agreement as the investors right to require your company to purchase the investors share according to predetermined criteria. A 'call' is the reverse, i.e. the right of the company to purchase the investors ownership. There are many forms of such provision, e.g. use of the Price : Earnings (PE) ratio. The investor will calculate an earnings per share, take an average PE ratio for companies in a similar sector (good sources of this information are the *Financial Times*, *Wall Street Journal* and *Genetic Engineering News*) and multiply the two to give a price per share. This will work if the company is in a relatively mature state with high earnings. However, a biotechnology start up may have high R & D expenditure in the early days and it may take several years for profits and hence earnings to accumulate. Pre-tax earnings are also not likely to give a true valuation since biotechnology companies often have to pay heavy salaries or undertake expensive marketing campaigns. Here a valuation based upon a percentage of sales is often used but has certain problems, e.g. sales could be artificially pumped up after a short period due to an advertising campaign. Other methods will include use of multiples of cash flow or sales, a pre-arranged cash value, a calculated book value (likely to be low in the early stages) or an appraisal value based on a combination of the above and carried out by a stock broker or industry analyst.

☛ Reorganisation and rationalisation

This is regarded as a polite exit term for bankruptcy. It is the end of the road for the investor and may be a good time for you to cut your losses too.

☛ Selling shares to a new investor

This requires little explanation and can be mediated by either yourself, i.e. you find someone willing to buy from the first investor, or by the investors themselves. This could mark a transition from an early stage to a late stage venture capital investor. A particularly satisfying exit route would be sale of shares to a corporate partner. For example, as a result of your biotechnological innovations your company may have arranged an R & D programme with a larger corporate sponsor and as such the latter may have a considerable strategic interest. It may well be that they wish to take a stake in your business (whilst watching to see how it develops) and an appropriate route may be for the investor to sell to the corporate sponsor.

☛ Liquidation

This is a frequent exit route. It will involve selling the assets, removing cash and generally dissolving the investment. Very much a damage limitation exercise and a failure.

Financing expansion

☛ Sale of products or services

If your company is already trading then you may be able to finance expansion through internally generated funding. Whilst sales of products and licence royalties will impact upon your cash flow, a substantial up-front licence fee or advance upon royalties for use of your IP in a defined manner could have significant effect. This may be applicable where you have protected a new process, such as a new method for protein purification. If a suitable licence fee structure can be negotiated, then this method of finance has considerable advantages, namely a lack of debt and no requirement to relinquish control. However, this may not always be possible since for example, your IP position may not be amenable to such an approach. Other sources of finance then become more attractive.

☛ Debt finance

For the small start up or a specific product development programme the most common form of finance is a secured loan from a bank. This means that you will need to find some personal or fixed asset to guarantee the loan, should you lose the money. We have already discussed bank loans and overdrafts, hire purchase and invoice purchasing for the more developed business, but what do you do if you do not have enough security? Many academics would not be prepared to throw their careers behind a business idea, by re-mortgaging the house and sacrificing safe permanent positions. Fortunately, the government has come up with a loan guarantee scheme for situations where there is insufficient security. Under these arrangements, the government will guarantee 70% of any loan up to £100 000, the proposition for which must be considered viable by a major bank. It must be a proposition that the bank finds an acceptable business proposition but would not normally approve for reasons of insufficient security. A wide variety of small business enterprises are eligible for loans that can be for between two and seven years, possibly including a capital repayment holiday. Application for large loans are referred to the Department of Employment for approval, but smaller loans can be dealt with by the bank directly.

☛ An equity injection

This will be available to finance any aspect of the business, providing the use of funds is as agreed ('signed off') in the business plan. There are primarily three sources of relevance here: (a) venture capital funds; (b) corporate investment; and (c) private investors.

Venture funds exist to cover a wide range of requirements. These are sometimes independent (some are quoted on the stock exchange) and sometimes spawned from larger financial institutions (captive and semi-captive funds). The largest fund in the UK is that of 3i (Investors In Industry) an independent venture fund owned originally by the Bank of England and major clearing banks, but which is now quoted on the stock exchange. There are international, national and regional funds covering various rounds of financing from start up (seed capital funds), to expansion, acquisition and turnaround (a term for rescue of a failing company). In addition, there are funds which specialise in particular sectors, including biotechnology. The venture capitalist exchanges funding in return for an agreed amount of share capital (equity) in your business. Unlike the bank financing route, the investment is not secured and the risks are as for the other shareholders. Likewise, if successful, there is a participation in selling the company, in which the investment is realised. In order to reduce the high level of risk, venture capital institutions will often work with others to provide the package of finance that is required. This is referred to as **syndication** and will have some advantages for you, i.e. the shares that you exchange for cash will be held by several investors, thus diluting any effect they may have on the decision making processes. The specialisation of venture funds will work in your favour since first and foremost you will be in the presence of those who have a good understanding of your industry. Remember, the emergence of the biotechnology industry in the US is largely due to the shrewd investment policy of the venture community.

The role of the company promoter

If you have a new idea for a start-up company, there will obviously be a need to engender interest in the company in order to raise the requisite capital. In legal parlance, the person who raises interest on behalf of the company is termed the 'promoter'. This will include forming the company, finding directors and shareholders, negotiating business contracts, etc. The activities of a promoter are subject to company law in a similar manner to that of directors, which is generally aimed towards the prevention of sharp operators making a quick turnover. The courts have decided that the promoter has a fiduciary relationship with the fledgling company. The promoter must give full disclosure of his interest in any transaction and the latter must be approved by the board of the company. The laws relating to the behaviour of a promoter are complex and if you are considering entering into such a relationship, you are advised to thoroughly investigate the position beforehand.

Venturing into share issues

In raising venture capital, unless you have done it before, you are likely to be in the hands of your financial advisor. The venture capital negotiators are hard nosed commercial beings who will be looking for the best deal for themselves that they can, consistent with their own investment policies. However, they are conscious of the need to leave the entrepreneur with a lot of incentives, otherwise there may be a feeling of working extremely hard for not enough personal benefit and the entrepreneur may thus expend less time and effort. The foregoing may seem obvious, but it will be worth looking in detail at the consequences of the share deal you are offered. In addition, a few 'what if' calculations may be wise before you finally agree to a deal. It is most likely that the venture capital company will propose a structure for the financial arrangements which will include the debt : equity ratio, suggestions for the entrepreneurs commitment and within the equity arrangements, a share structure. Always contact an independent financial advisor for discussion of your particular circumstance.

☛ Ordinary shares

This is the simplest form of share and is rarely used for this type of financing. For example, if the entrepreneur is prepared to give up 20% of the company in exchange for a £200 000 investment.

	No. of shares at £1 each	Ownership	Outlay
Entrepreneur	80	80%	£80
Investor	20	20%	£200 000[a]

[a] Implies a premium of £9 999 per share.

At the time of investment you can expect to pay an arrangement fee to the investor (i.e. for somebody's time in coming up with the numbers), professional fees (solicitor/accountant) of between £5000 and £10 000 and a 1% stamp duty if there has been any transfer of property rights. In addition, it is likely that the venture company will appoint a director, who will then receive a fee of between £3K and £10K plus of course, reimbursement of expenses.

☛ Dividend payments

Some venture capital companies will require that a certain percentage of profits (between 25 and 35%) be declared as an annual dividend. However, these can only be paid when there is a cumulative company profit. Dividends such as this will be subject to advance corporation tax (ACT) at 25%. Thus if your company declares a gross profit of £400K, the (say) 25% is declared as dividend, i.e. £100K. Of this, £25K will be payable to the Inland Revenue as ACT, thus leaving £75K to split between the entrepreneur and investor in an 80 : 20 ratio, i.e. £60K and £15K. This will be treated as gross income for personal tax purposes, but the Inland Revenue will grant a credit of the ACT that has already been paid. The investor will receive 20%, i.e. £15K that will be taxed according to status. Clearly, such a mechanism will allow you to derive substantial benefit from the success of your company, in addition to your salary.

☛ Capital gains tax

Let us assume that after four years, the investor wishes to sell the 20% for cash and that there is a buyer willing to pay £500K. To calculate the net capital gain we must deduct the original input (£200K) and an allowance for real gain, that may not be apparent from using an inflation rate alone. In practice, the retail price index is used. Let us assume this to be 5% increase for each of the four years. Hence the indexation allowance is $[(1 + 0.05)^4 - 1] \times 200 = £43\,000$. Thus the net gain is $£4\,000\,000 - £200\,000 - £43\,000 = £3\,757\,000$. Of this, capital gains tax (CGT) will be payable at 35% for the investor (20% share) and 40% for the individual (80% share), i.e. £263 000 and £1 202 000 respectively. This should leave you, the entrepreneur with £1 803 360 and the investor with £488 410. The added value will come in terms of fixed assets and equipment (e.g. well equipped laboratories that you own), highly trained staff and a significant base of intellectual property, to help ensure a future income stream. This example is an over simplification of reality, but it does illustrate some relevant principles.

☛ A mixture of ordinary and preference shares

The use of preference share is a mechanism by which the anomaly of ownership, i.e. an entrepreneur with 80% of the company and 0.04% of the investment can be evened out somewhat. Hence preference share holders are entitled to a dividend before the ordinary shareholders, i.e. if there is money in the pot the preference shareholders get it first and in any sale of the company in the future, the preference share holders are paid first. Since control of a company is determined by the amount of ordinary share capital, the issue of preference shares will not dilute your overall level of influence upon its management. Similarly, with cumulative shares all dividends that would have been due over a period (but have not been paid because of insufficient funds) will accumulate and be paid up to date before the ordinary shareholders receive funds. The amount of dividend (sometimes referred to as a coupon) will usually be predetermined, i.e. a fixed percentage. Participating shares allow the holder to participate in a dividend that is based on the level of company profits, which may be expressed as a fixed percentage of the net profits or of the net dividend. Convertible shares represent preference share capital that can be converted to ordinary shares, e.g. upon sale of the company or if there has been some reason to suspect default on the investment contract. Redeemable shares are again preference share capital that is repaid at a given time period, with a pre-agreed repayment schedule, say five years after investment or upon initial public offering (IPO) or sale. Such shares will also carry a requirement for repayment if there has been a **default** in the contract at the time of investment, e.g. if relevant information was withheld. Sometimes a mixture of debt and equity will be appropriate, e.g. a bank loan and ordinary shares. There may be some tax advantages for you to do this, e.g. tax will not be payable on monies used to pay bank loan interest. Your use of loans instead of preference shares will mean that in the event of failure, loans will be paid off first and thus there is less chance of your personal guarantees being exercised and on a more positive note, you will be entitled to a financial return, since there will be a reduced (or no) obligation to preference shareholders. Later stage investments may use a mixture of ordinary and preference shares together with loans or debt. This depends entirely upon the financial package that can be negotiated between the investor and yourself. As an extra performance incentive, the investor may offer you (the entrepreneur) some share options at the time of investment that can be converted to ordinary shares upon reaching particular performance levels, e.g. profit expectations.

Sources of risk finance

☛ Venture capital

If your business venture has a considerable risk element, then it is unlikely that you will be able to borrow money from a bank and other sources of finance to which you may have access might not be sufficient for your needs. At this point you may like to consider raising a form of risk finance and the form most likely to be applicable is venture capital. This finance is likely to be as an equity investment in the company such that the investment can grow as the business grows (Fig. 8.1). The venture capital route is that of a medium- to long-term investment of both time and money on everyones part, with the view to the building of a significant business. There are various stages of company development and therefore, a series of stages at which investment may be made. The venture capitalists often specialise in these particular stages and in so doing build up considerable expertise in the principles and problems associated with said stages. The earlier stages often hold higher risks, with the prospect of higher rewards via a larger share ownership and the advantages of creating value by dilution.

In addition, there will be a competent evaluation of your business proposal with hopefully some constructive advice. However, in order to be accepted for this due

Fund	Function
Seed	Development of a business concept, prototype production, market research
Start up	Development of a product, marketing. For companies not yet trading
Early stage	Initiation of manufacture and sales. For companies not yet trading
Expansion	To increase production, provide working capital, do market research and develop new products
Secondary purchase	Purchase of shares from your own investors or in another company
Management buy in	Allows external management to buy into the business
Management buy out	Allows encumbent management to buy the business

Fig. 8.1. Stages of venture investment.

diligence process, your proposal will have already needed to pass several hurdles. How then, do you decide how to approach? The best route is to obtain the free directory of members of the British Venture Capital Association which gives a full listing of investor details, including the technical and regional specialisation together with the maximum and minimum investment. It is suggested that you make a list of companies that you think may be suitable and then approach your financial advisor or accountant to filter this list down to two or three companies for final approach. A 'scud' missile attack on investors is not likely to provide the required result and a more directed (cruise missile) approach is likely to bring you together with an investor who appreciates being chosen with thought. Indeed, you should be prepared for a coherent answer to the first question the investor is likely to ask you, i.e. 'why have you come to us?'.

☞ *The approach*

In the first instance send a short (3 to 4 line) covering letter and the executive summary from your business plan. Preferably, either telephone beforehand to briefly explain and leave the investor expecting the proposal or try to arrange a personal introduction via your financial advisor. It is in these cases that you will find an association with an accountancy firm to be a considerable asset.

Two days after the expected day of receipt of the letter and if you have not had a response, make a follow-up telephone call. If they have not read it or indeed have lost it, fax another copy through immediately so that the investor is expecting the document. Try not to be too pushy or enthusiastic (this can be irritating) but calm, professional and persistent until you get a response. If the response is positive, then promptly provide the information requested. If negative, find out the reasons for rejection since this may affect how you approach the next investor. Remember to make sure the executive summary indicates the potential for sustained growth and the ability of the management to carry through the business plan. The investor will be looking to see if the rewards justify the risk and whether the rewards are consistent with their own investment criteria.

☞ *The initial negotiation*

At the next stage you will be invited to meet the investor who will be evaluating both yourself and the proposed management team. In some respects it could be better to seek advice from the investor as to the structure of the management team and before it has crystallised, since you need to ensure that the launch pad is built upon solid ground. The investor will be looking to see if you can 'sell' the business plan, if you understand the market, if you have the relevant information systems, etc., i.e. you will be interviewed on your business plan. This is your opportunity to build a good working relationship with the venture capital organisation.

The next stage is to commence commercial negotiations. One of the investors will wish to be the **lead investor** and they will be competing amongst themselves for this honour. This is your opportunity to negotiate commercial advantage for your company and it is recommended that you concurrently negotiate with around three investors hoping that one may take the lead role. It may then be that the lead investor will syndicate the deal to others in order to reduce the risk.

The extent of investment and equity holding requested together can be used to calculate the value that the investor has placed on your company, e.g. a 10% holding for £50000 means a value of £500000. Independent negotiation with three investors will thus give you an idea of the value of the business to them and this figure will be calculated to be in their favour, not yours. This means that there will be a tendency to undervalue the business, since the investor will be looking for maximal shareholding for minimal financial exposure. By definition, biotechnology start ups will not be quoted on the stock exchange and thus do not have a track record from which to calculate value on the basis of share price (the price : earnings ratio is often used). Factors in your favour from having an unquoted company are: (a) you can engender competition amongst venture capitalists as already discussed; (b) the business will be in a trendy area (one only has to look at the explosion in 'apoptosis' related biotechnology start ups); and (c) there will be a considerably higher than normal growth rate and hence profit stream. These factors will tend to increase the value of your business. However, besides the short track record (if any) the value of the company will be reduced relative to quoted companies since the shares cannot be traded, the cost of making and monitoring the investment will be higher and there are higher risks associated with a lack of diversification.

The financial rewards will have been calculated on the basis of an IRR but the absolute figure expected will vary according to sector and stage. For early stage biotechnology businesses this is likely to be high (e.g. 30–40% IRR required) since there are many failures and the high returns are needed to offset the losses. Needless to say, some commitment will be required from yourself and either the equivalent of one years salary or 10% of the capital to be invested in the company will perhaps be considered reasonable. Remember, you will probably have a shareholding in excess of 10% so the investment will not be *pari passu*. With regard to your own position, one tip is to ask for stage financing, i.e. rather than take the equity finance all at once and at the beginning, rounds of finance are provided upon meeting particular milestones that have been pre-agreed. In this way the risk to the investor is reduced and you should be able to negotiate a higher equity share for yourself on this basis. Similarly, you should negotiate with the investor on whether: (a) there is to be a fee for completing the deal and how much this will be; (b) who is to pay for appraisal of your project (due diligence); (c) the extent of indemnities and warranty that the company directors provide; (d) if the investor is to have a seat on the board; and (e) the number of votes that will be associated with the shares of the investor.

Upon satisfactory completion of these negotiations you will receive an 'offer' from the investor, which is not a legally binding document, serving only to summarise the position to date and indicate that the investor is taking the proposal seriously. The next step is for the investor to carry out a full project appraisal, which is known as the due diligence process. The proposal must be assessed with regard to technical aspects and financial robustness, a process which will involve outside consultants, scientific peers, interviews with potential customers as to whether they would be interested in such a product, an accountants appraisal of your profit forecasts, bank facilities, management information systems, etc.

In addition, the investor will take references on both your company and the individuals in the management team. This due diligence should be sufficient to release any skeletons from the closet and you should be prepared to finally lay to rest any of your potential investors concerns, should they wish to take the project further.

The final stage is one of completion, whereby Heads of Agreement are followed by a series of legally binding completion documents as follows:

☞ *The shareholders or subscription agreement*

This contains the terms under which the investment will take place, warranties and indemnities of the existing shareholders and any obligations that the management may have with respect to the investor. The warranties provide proof of the information provided, to which you will be held liable for inaccuracies. If indemnities are included, the directors/shareholders agree to reimburse the investor for specified accounts in the event of business failure. Also included should be service contracts for the directors and management and a disclosure agreement. The latter contains the information you have provided to the investor for decision making purposes. In addition, you will probably require new memoranda and articles of association to document the share rights that will be attached to the investment. There will be costs associated with raising the finance, being around 5% of the equity raised, some of which will be a fee for the investor. This latter is negotiable and furthermore, you should make sure that it is agreed who pays before any work begins, preferably with a written agreement. This then allows for payment of costs should relationships break down. A document of some importance is the disclosure letter. This contains all of the key information that you have previously disclosed to the investor and must contain that information which was used to assess the investment; as such this serves to limit the warranties and indemnities to any points not disclosed previously to the investor. The completion event consists of several hours of document ratification in the presence of lawyers, followed by the actual investment of cash under the terms agreed.

☞ *Post investment monitoring by the investor*

Investors may either play an active or inactive role in the management of the business and can make a considerable contribution to it. However, there is often a tendency to regard the investor as an investment policeman. In any event, the investor will expect to receive both the management accounts on a monthly basis and the minutes of board meetings. If the conditions of the investment require appointment of a representative of the investor to the board, then this person will no doubt participate in company decisions or failing this, the investor may retain the right to be involved in major decisions which affect the business of the company. In particular, the investor will be attempting to protect the investment by helping to identify business danger signs, for example, increasing fixed costs, over-trading, inacccurate management information, high staff turnover, over dependnce upon a low number of suppliers and customers, and loss of credit control.

☞ *Relationships between yourself and the investor*

Entrepreneurs such as yourself and the investors could have a fraught relationship. At the time of investment you will be glad of the money and the garden will look rosy. As time passes, so the investor may feel that if only he/she was running the business it would do so much better. And you may feel that you have done all the work in making the business a success whilst the investor has done nothing. If results are not forthcoming so the relationship may become even more strained and so at any time the investor may need to try to 'save' the investment. Even if things are going well and you have met the investment criteria, the investor may wish to exit to use those funds for other new investments. Either way, do realise that investors can 'sell up' at any time. Having said that, most equity investments are for a long period of time, say between three and seven years, since this is the time which is most likely to accrue before investment criteria (e.g. a 35% IRR) are met.

☞ *Venture capital in the UK and Europe*

The representative body for the UK venture capital industry is the British Venture Capital Association (BVCA) and with greater than 200 members, can help make the connections between yourself and sources of capital. It publishes several guides, the most useful of which is a book that lists all of its members, giving addresses, contact names and most importantly the members that are most likely to invest in a particular field. There are only a handful of investors who are serious about biotechnology in the UK, there still being a reticence and short-term attitude regarding investments in this sector in the UK. In the initial stage, you may require only a small amount of capital and as such, the Midland Bank maintains an excellent database of potential investors who are prepared to provide relatively small amounts. There are similar guides available for sources of venture capital in Europe and in particular, Venture Capital Report publishes a respected volume on the subject. In addition, the EC provides information on seed capital funds that invest according to criteria set by the Commission and provides support for projects that arise from RTD programmes. This 'Eurotech'capital is administered through a network of particular venture capital companies.

☞ *The investors view of the proposed investment*

We have now discussed the basic requirement from your point of view, as a recipient of the investment especially with regard to the basics of business plan preparation. However, it will be advantageous to discuss the investment from the investors point of view and to give a perspective in the standards expected. You will of course be expected to defend the business plan, but this is only a small part of the interview. The decision to invest will be on a number of criteria, mostly deriving from the investor's analysis of you as a leader in your new company. This will take the form of a personal interview that explores many aspects of your personality, motivation and ability to manage in a multidisciplinary process. Let us begin by trying to discover if there is such a person as a typical entrepreneur and of course you may like to judge if you have some of the necessary characteristics. In essence, the text will be assuming that you are the entrepreneur, perhaps looking for the first investment in a start up biotechnology company.

There have been many studies of entrepreneurial character by social scientists and no doubt many an MBA theses on the subject. One should perhaps consider four sets of characteristics: (a) behaviour patterns (realism, determination, resourcefulness); (b) mental characteristics (i.e. need for self-actualisation, the extent of desire for power, control of ones own destiny); (c) physical prowess (stamina, oral and written communication skills; and (d) integrity (honesty, desire for fair play).

Behaviour patterns The investor will be looking for an entrepreneur who is determined to succeed and can set a goal and leave no stone unturned until that goal is reached. In addition the person should show resourcefulness, i.e. be able to find solutions to problems for which there is no obvious solution or resource. This should sit well with those of a scientific disposition since this is supposed to confer both creativity and inquisitiveness. However, there is a great deal of difference between the creative input required to solve business problems and those that focus on a specialist technical area. The investor will be wary of someone who has an excessive sense of urgency; the tendency to take on too much and not really accomplish anything. Make sure you are on time for meetings, prepared and refreshed, since there are few better indications of an overstretched individual than the contrary to the former. The investor will be looking to see if you are in touch with the reality of your situation, so expect to be challenged on the financial projections. It is easy to believe your own 'hype' eventually, but try to differentiate between evidence based projections and wishful fantasy.

Mental characteristics The investor will be taking a view on your intelligence level via general conversation and not just upon technical rhetoric. It is relatively easy to bamboozle a lay person in a scientific subject and the shrewd investor will not see this as an indication of mental prowess. Good entrepreneurs have a passion for achievement and being the best; they are often very competitive people. This is relatively easy for the investor to determine since your past history can be investigated with regard to your achievements.

Entrepreneurs also have a desire for autonomy and independence, tending to be a member of few groups or clubs, not having a mentor and generally having poor interpersonal skills. There is also the rather unpleasant characteristic of entrepreneurs in that they need to control others. Unless held in check, this can have disastrous consequences on the motivation of those around them and can severely compromise the chances of success. Entrepreneurs have little time for fate, believing that they are in control of their own destiny, not withstanding opportunism. In general, they are confident people often to the extent of arrogance. Contrary to expectations, entrepreneurs tend to take moderate risks, in contrast to the low-risk strategies of bankers and the high risks taken by gamblers. Each of us has a risk profile of what we consider to be acceptable risk. In general, it is not easy to translate from a 'research-risk' mentality to an investment and business risk appraisal, so to avoid embarrassment, do not do so until you have sought proper advice. Entrepreneurs are also characterised by having mental stamina, i.e. an ability to focus and concentrate on a given problem for a long period of time. If the interview with the investor appears to be protracted, then this is probably what is being put to the test.

Physical prowess The entrepreneur is always energetic and should turn a large amount of time and energy towards the business. Do not be surprised if the investor requests a life insurance policy on you or checks if you are working, by making early morning and late night phone calls to you. In addition, the entrepreneur should have social persuasiveness and charisma, i.e. the oral skills necessary to convince others of the validity of a case and considerable charm in general conversation. In addition, you are also looking for someone with a clarity of purpose and eloquence in the written word; this is necessary for the preparation of board reports, the business plan and monthly management reports. If any of these are ambiguous or badly written, they are next to useless.

Integrity There is no doubt that the investor will be looking for a high degree of integrity, honesty and straightforwardness in interactions. There will be a belief that at sometime during the relationship the investor will be finessed of financial advantage should the above not be true. The best relationship with the investor is one of partnership and hence concurrent goals, ideally to build a business and realise the investment in both time and money by sale of the business. The investor will be looking for someone with an air of detachment and not a deep seated personal interest that may preclude an appropriate exit route.

In an interview situation it is extremely difficult to assess honesty. You may be tripped by repeat questions or be seen to avoid a pertinent question, but aside from this, the investor will be forced to seek alternative impressions, e.g. from personal reference, credit records, etc. There is of course a class of individual who will exaggerate a position during negotiations, not give bona fide explanations and generally lie about business issues. Indeed, I have come across two individuals who would, to differing degrees, say absolutely anything to win the smallest argument even amongst colleagues. Both were capable of tripping themselves in the next sentence, but of course the deceptions were not always easy to detect. The inconsistencies therein eventually and inevitably come to light, resulting in a total loss of credibility, i.e. it was subsequently difficult to believe anything that these individuals said. Try not to be like this; it can only lead to a destruction of relationships, a lack of progress and a considerable number of lost opportunities.

To summarise, the investor will be collating this information in order to ascertain: (a) how you are likely to treat the venture capital company; (b) if your personality is suitable for the industry sector you wish to enter (there are not many precedents amongst UK academics); and (c) will this personality make the businesss itself successful.

☞ *The essential entrepreneur*

The venture capitalist will be looking for individuals with the ability to concentrate their efforts over a long period of time, believing that success is a function of perspiration (90%) and inspiration (10%). If a small company is to succeed, then long hours are required to compensate for under resourcing, notably the lack of staff. Entrepreneurs must also be comfortable in risky situations, whereby they have the ability to evaluate the circumstances and act in an appropriate manner; there will be many such situations in the emerging biotechnology company. Commensurate with this is the ability to pay attention to details, not perhaps in a pedantic manner, but in sufficent depth to make correct decision making possible. Excessive preoccupation with all minutae will divert one's efforts from the high value added exercises, such as new business development.

The budding entrepreneur must also be able to articulate ideas and strategy, to the extent that financiers, bank managers, employees, government representat-ives, etc. are convinced of their ability to bring the venture to a successful conclusion. This leads us onto the concept of a compatible personality, since the investor will be looking for someone with whom they can work with over a long period of time. Whilst this is often a matter of personal chemistry, obnoxious people are rarely pleasant to interact with.

There should also be a firm understanding of the marketplace and as the leader of the company you should both know and research your potential market thoroughly. Having been in a research laboratory is not sufficient and the investor will be looking for some evidence of commercial activity in the chosen arena. This can be difficult for someone wishing to step into the business world and, until you have realised the parameters, it is wise to have the partnership of an experienced business person. Investors will be looking for relevant track records amongst the individuals concerned and in a sense, it does not matter if a previous venture was not successful; hopefully you will have learnt from the experience and will be unlikely to make the same mistake. The investor will also expect to see strong leadership skills and a significant reputation in the field of interest. Furthermore, you should expect to be able to provide references as to your own character and abilities.

☞ *The biotechnology business environment and venture capital*

In contrast to the US, in the UK there have been relatively few partnerships between biotechnology start-up companies and corporate bodies, so there is a lack of reassurance for the investors. Concurrent with this, profit margins in Europe have been smaller than in the US due in part to the fragmented national markets, i.e. the costs incurred in dealing with regulatory affairs and separate marketing strategies. However, the climate is changing and European markets are now integrating in a number of sectors. Furthermore, governmental and public concern regarding healthcare costs in the US are further likely to reduce profits, making the European markets seem more attractive to investors.

The most successful biotechnology companies in terms of sales revenue have been characterised by their ability to raise large quantities of cash. Many companies are under capitalized and have not correctly estimated either the amount of money required to take a new drug to market ($100m according to some estimates) or the high consumption of cash by the research and development process. The US norm has been to raise as much money as possible for a new venture and search for corporate collaborators. European companies have set their sights on similar goals but have tended to raise as much money as they think necessary, which has perhaps caused a degree of cash starvation. This then increases the risk for the investors who are likely to find themselves propping up non-profitable ventures, with the company continually on the back foot. The management then spends valuable time in seeking further rounds of finance instead of allocating adequate time to managing the business and concentrating on product development and marketing. Such companies are often referred to as the 'walking dead', since there will have been a considerable amount of money invested in research, but a dearth of resources remaining for personnel and facility expansion, new product development and marketing. Inevitably, the cycle continues and the investors require a return on the investment, which can often lead to compromise in research into new products and an increase in revenue generation by, for example, contract research. Overall, these and similar events will reduce the attractiveness of the business as regards long-term earnings potential and hence it becomes extremely difficult to raise further capital. Ideally, the best option would be to raise a considerable quantity of cash at the outset, since a large capital base goes a long way to reducing the risk in an investment. The experiences of the 1980s may make this more difficult in today's investment climate. An alternative would be to lower the expectations and concentrate upon building a company with limited dependence upon outside capital. Obviously, we cannot then include a large research element in the equation and the focus is more likely to be upon products for research use and contract research, the latter perhaps being undertaken in a development laboratory. The scaling down of the projections may make you less attractive to the venture capital markets but will make the proposition seem more appealing to private investors. In this way, the owner (probably yourself) holds onto the equity and is in a good position to make significant financial gains even if the company remains unlisted on the stock-market and is only modestly successful.

The ideal exit route for the venture capitalist is the introduction of the company into the stock market. Large amounts of finance can be raised by this route and the company then assumes a more mature culture with a decreased dependence upon the high risk research and development divisions. The mixed fortunes of biotechnology investments has led to a rationalisation of venture funds in the US, with the larger funds now increasing in size, whilst the effects of the occasional bad investment have taken a toll on the smaller funds. The rules and possibilities for company flotation are seen by many to be prohibitive for young companies in Europe, although some moves have been made to alleviate this problem. In addition, European business tends to cater for large quoted companies and there is a shortage of public markets for flotation of small companies. This contrasts to the US where flotation on the stock market is more common and somewhat easier to achieve. Thus investors have a well-tried and tested exit route and hence mode of realization of their investment, helping to convince them of the shrewdness of the venture.

Following on from the private placement route, a third method of financing a research based company is via corporate investment. It may be that a company is extremely good at research and development, yet lacks the necessary downstream skills. For example, building the sales marketing network to fully exploit a new pharmaceutical product would be quite prohibitive, especially when one considers that you may cut into existing markets that a larger player(s) may wish to protect.

Venture capital is but one route to financing your venture. Two other routes that should be investigated are corporate venturing and the use of 'business angels'.

Corporate venturing

Corporate venturing involves the purchase of an equity stake in a small company by a larger company. For example, a major customer may vertically integrate with an innovative supplier. This arrangement is becoming more common in biotechnology, especially as some of the smaller companies are now bringing drugs through to the clinical trial stage and require a marketing network. Others have proven to be consistently innovative in a particular therapeutic area. In some cases senior executives of the pharmaceutical industry have left to form their own companies, essentially working for themselves as opposed to the larger corporation. It is expected that as biotechnology companies mature they will become more attractive to larger partners and more of these relationships will emerge. Indeed, investing in young companies is one means by which multinationals can aquire innovative products, access new technology and take advantage of an entrepreneurial culture whilst dealing with increasing financial pressures on the home front. There are three prevalent stages in corporate venturing. (1) The larger company takes an equity stake in the innovative company in return for first option on specified elements of the research. (2) A development programme will be delineated, to include milestones, royalty payments and terms of licensing. This will include the rights to marketing in specified therapeutic or geographical areas. (3) Provisions must be made for the long-term relationship, which may include either pre-emption rights over the remaining equity or the exit routes that are open to the corporate body.

In considering such a venture, there are both advantages and disadvantages to the small company, together with a number of global issues:

1. Make sure that the project is clearly defined from the outset, i.e. resources, personnel, project milestones, performance criteria, beneficiaries and financial issues, e.g. the rates of return that will be required.
2. Make sure that both partners are fully commited to the project from an internal point of view and that the interface between them is well managed. There will be considerable cultural differences; a balance must be sought between bureaucracy and entrepreneurial vision.
3. Evaluate scenarios of failure and lay contingencies for them. This is particulary relevant while multinational companies are still evaluating the wisdom of corporate venturing as an activity.

Corporate venturing is increasing in popularity and the investment is usually managed in one of two ways: (a) the use of a majority owned venture capital fund that operates outside the company framework but may use company staff (e.g. Elf Aquitaine, Sandoz and Smith Kline Beecham have funds called Innovelf, Avalon and SR One respectively); and (b) the use of in-house teams that make direct investments. From the small company point of view, the advantages are: (a) a ready source of cash to help maintain viability; (b) the opportunity to create a bond with a large partner, thus offering an exit route; (c) the probability of a larger equity stake remaining with the entrpreneur than if dealing with a venture capitalist; (d) access to databases, support services; and (e) the ability to continue doing what you are good at. The disadvantages are that: (a) you will be required to report to the sponsor in a way that is acceptable to them, i.e. there will be an increase in bureaucracy; (b) you may be the recipient of 'interference' from the sponsor's scientific staff; (c) contract negotiations can be protracted, when you really need to get on with the job of managing the company; and (d) you can easily find yourself facing superior forces with respect to professional advisors.

The company will also be facing some advantages and disadvantages with respect to investment in your company and some of these points will be presented here. These are given for the purposes of understanding your potential partner's point of view that may be useful in negotiations. Equity is purchased in a deal that trades the acceptable control systems for innovation and lateral thinking in a particular research area. Thus tangible becomes intangible; your attitude and professionalism are crucial here, as it is essential to make the sponsor feel comfortable. Make sure you have project reporting sytems in place and a vibrant and competent organisation on show in order to subtly enhance your credibility. The sponsor is gaining access to research that does not have a full corporate overhead allocation, since besides avoiding the in-house research costs, the small company is likely to be run along more frugal lines. However, the corporate sponsor will be wary of being overly dependent upon outside technology and also of the possibility of breaches of secrecy. The hidden agenda will be one of full aquisition if the venture is successful.

☛ Business angels

Private investors (business angels) are usually individuals with capital to invest in what they believe to be a company with growth potential. This type of investment may be referred to as informal equity capital and can often be obtained upon more favourable terms than the equivalent venture capital. Private investors can be a breed unto themselves, although to be fair they are usually investing their own money, which does shift the motivation spectrum somewhat. Thus they often wish to play an active role in the development of the company, by taking a board position, working part-time or acting as a consultant. As with other investors, business angels will be looking for financial realisation and hence an exit route. The distinct advantage of this route to finance is that the investor is likely to make a rapid decision, often being dependent upon their own expertise rather than an extensive due diligence process. However, the disadvantages are that the business angels may be unwilling or unable to provide subsequent rounds of finance and will often expect to influence strategy, which could lead to disagreements. Selected local enterprise agencies offer a partnering service via the LINC (Local Investment Networking Company) system. In addition, various publications are available such as 'Sources of Business Angel Capital', which is available from the BVCA, and the Venture Capital Report (VCR). Private investors subscribe to VCR and use it as a source of information concerning a wide range of projects that require investment in the £20 000 to £2m range. In order to feature your project, you can either invite a one page advertisement for a one-off fee or have VCR write an article about you and your project, to include a publication fee and a success fee. This may indeed be a good route to approaching private investors. Similar services are offered by 'Techinvest' and 'The Capital Exchange'.

Government policy has changed towards the business expansion scheme (BES), which had, until the 1993 budget, been an excellent vehicle for private investors to support small- and medium-sized enterprises. Instead we now have the Enterprise Investment Scheme (EIS) whereby 20% tax relief can be obtained on a maximum investment of £100 000 in companies that are not quoted on the stock exchange. There are a number of tax and other benefits of such schemes to the investor that could make your business an attractive proposition. However, care should be given to how the investor will realise the cash input and to the extent of control exerted by said investor, particularly with regard to directorships. In this regard you should pay particular attention to your memoranda of association and seek professional advice. Similarly, 1995 heralded the introduction of Venture Capital Trusts whereby investors can subscribe to high-risk investment trusts whilst taking advantage of various tax concessions. The fund managers will not be able to invest more than £1m in each business in any given year and the business must be valued at less than £10m, which should make more capital available for small entrepreneurial businesses.

Creating value

☛ The likely route of business development and share dilution

At all points in the development of your business, potential investors will be looking for an appraisal of its value. All that can be expected is that this value is 'reasonable' for the business, when comparing its goals with the time-frames involved. An undervalued business is likely to reflect inherent weaknesses, perhaps in the product line or management, whilst an overvalued business is not likely to maximise the return to the investors. This is because the business will be close to its target value and thus the margin for investment return will be reduced. With many biotechnology businesses, the valuation is highly subjective, since they are often loss making due to a high R & D expenditure in the early years and a lack of product sales. Often the criteria used are: (a) the number of products in clinical trials, together with their stage and therapeutic potential; and (b) the number of collaborations with major companies, e.g. drug development programmes with the pharmaceutical sector. The valuation would be expected to progressively increase as the business develops and again, the rate of increase will be monitored. Too slow and one might be tempted to assume a poor product profile or an inappropriate management. In the initial formation of the business you, as the founder, will take 100% of the equity in the business. It is then likely that you will require additional funds from an early stage, perhaps involving private investors and finance from a venture capital fund that specialises in such a stage. By this time, the company will be established in terms of premises, staff, running budgets, etc. and will hopefully have secured an IP position. The action of raising the extra funds will have resulted in the first dilution of the owners shareholding but the overall value will have increased. The latter follows since otherwise the investors would not have invested. The cycle will be repeated as your business plan unfolds over several years, with successive rounds of financing from increasingly larger funds. These are likely to include venture capital funds, specialist investment funds and institutional funds. At each financing round, the share value will increase in price with the magnitude of the increase being dependent on you being able to demonstrate progress during the intervening rounds, i.e. commercial success.

In order to satisfy the investors, the value of the shareholding must increase since otherwise there would have been very little point in making the investment. This means that when it comes to making a public offering of your company, there is a minimum share value below which it is unlikely that more shares, i.e. further dilution, will be permitted. The value of the share price will equilibrate with other companies on the market as value comparisons are made with companies in a similar position. Whilst this is difficult to assess in the UK, due to the currently small sample size, the share prices of US biotechnology companies can be compared within the sector. A good source of this information are the regular reports which are published in *Genetic Engineering News*.

☛ What are the criteria that contribute to a good biotechnology business?

This section is included in order for you to have a view as to the sort of company you could eventually build. The principles apply to most businesses in the biotechnology sector, although there will be exceptions.

☞ *Unmet medical need*

The ideal circumstance is to focus on diseases that represent unsatisfied markets and where there is low competition. This rather glib statement is the holy grail of the drug discovery, therapeutic and diagnostics arenas and there are a lot of groups searching for this competitive edge. The advantage that you will have is: (a) the specialist knowledge of a disease area or particular technology; and (b) the ability to move quickly and assume some risk, characteristics that are not necessarily applicable to large organisations. In the short term, one could focus on 'headline' business, such as drugs associated with either stimulation or inhibition of apoptosis, the emergence of multiple drug-resistant strains of tuberculosis, anti-HIV agents or systems for the early diagnosis of cancer. Perhaps one could focus on longer-term goals and tie the development of the company to the shift in demographics, i.e. the expected increase in older people and consequent increase in the diseases of ageing (e.g. diabetes, cancer and cardiovascular disorders). Since it can take up to 10 years to develop a drug and take it to market, (3 years in R & D and preclinical testing, 7 years in clinical trials), businesses that are based upon the diseases of ageing are likely to have a healthy future in 10 years time.

Innovative products will command high prices, for both commercial reasons (skimming the market to increase margins) and practical reasons (to cover the large cost of R & D programmes and to reinvest in new products). It is logical to assume therefore, that the companies that are most likely to succeed are likely to be the most innovative. The value of these companies will in part be reflected in their importance to the pharmaceutical companies with which they have developed alliances.

☞ *Corporate alliances*

As credibility is aquired through: (a) innovative products and services; (b) a developing financial and scientific reputation; and (c) demonstrable achievement of milestones, so your business will become attractive for corporate liaisons. This is likely to give access to additional sources of funding and perhaps even product development expertise.

☞ *Management*

Different stages of business development requires a different set of skills and your research abilities could either rapidly become redundant or be confined to new product development (important nevertheless). It is unfortunate that our educational system selects solely for a research training, and that those with the personality and aptitude in other areas fall by the wayside until they flourish in larger organisations at later points in their careers. There is not a shortage of research scientists in the UK, but for those with product development skills and in regulatory matters there is no training programme except for the experience of life. We badly need a generation of young entrepreneurial development and regulatory managers in our biotechnology industry.

There is no doubt that some of biotechnology's so-called 'failures' could be attributable to the poor design and management of clinical trials, thus delaying and/or preventing regulatory approval and introduction into the market. The potential investors will be aware of this issue; moreover, they will be aware of the large amounts of money that have been injected up to this point and hence their increased individual exposure. Given this, it is likely that they will be looking for a management team with developmental and regulatory experience and who have already taken a drug to market in the US. Therefore, this implies team members who have successfully taken a drug through the Federal Drug Administration

(FDA). Ideally you should be able to provide evidence of your company's ability to progress lead compounds through to registration in a rapid manner. Whilst this is not an immediate concern for the small entrepreneurial R & D start up, you should be aware of the issues and begin to lay contingency plans.

☞ *Running a tight ship*

One of the key elements that will demonstrate your ability to run the business is in your ability to control the costs, to derive and stick to realistic budgets and to correctly predict future expenditure. At first sight, you might consider that this is a way of showing the investors that you are using their money wisely and not wasting it on frivolous entertainment budgets, etc. However, the underlying reason is more profound; overspending will cause cash starvation and to keep the business going you will have to raise more money. This will require management time, which should be spent on other things. More to the point, if you are successful in raising this money, then there will be a resultant dilution of share ownership, a scenario that is not likely to please even the most tolerant of investors.

Whilst R & D is likely to be a big drain on resources, full scale clinical trials are likely to be more so. Hence if you reach this stage with one of the products of your biotechnology start ups, the investors will be looking for a tight rein on the clinical trial costs. Costs of such trials are likely to be somewhat reduced in the UK compared to the US.

☞ *Protection of both intellectual property and the product line*

If you are to maintain the prospects of high profit margins in a market with few competitive products, then you must ensure that the intellectual property which surrounds your product is well protected. Regulatory authorities are helping to speed some of these unprecedented products to the market by granting 'orphan drug' status, which has helped some of the younger companies considerably and reduces the risk to the investor. When considering your company profile, you must try to ensure access to a stream of new therapies and technologies; over dependence upon one product and one technology considerably increase the risk to the investor and you must try to reduce this (product line extension, new products using the existing technology and complementary technologies and products in related fields).

If you have or can develop rights to generic technologies, then this is a particularly advantageous position in which to be since it can provide the launch pad for a wide range of new products. You may also consider trying to reduce the time to clinical trial (the lead time); a large screening programme of chemical compounds that you have synthesised is likely to have a long lead time, a more precisely designed antisense, ribozyme or triple helix is likely to be less so. The longer the lead time, the more risk there will be to the potential investor. To put it another way, the investment risk will be lowered the closer a product is to market.

☛ The dangers of creating false expectations

There is often a tendency to 'oversell' a potential investment in the belief that this will increase the chances of obtaining the finance. This should be resisted. The investor will appreciate your enthusiasm, but will also be impressed by your pragmatism. If you extend reality beyond the normal bounds, the chances are that you will be held to your statements, especially when the going gets tough.

☛ What will happen if things go wrong?

You will know when your company is going well and conversely, when it is not. So will the investors. However, the entrepreneur is likely to paint a rosy picture even in adversity, trying not to rock the boat in heavy seas. The investment community has seen this many times and your company accounts, behaviour patterns and feedback from a whole host of individuals is likely to portray the correct picture. However, the investor will not be looking for problem investments and is aware that quick financial fixes (e.g. to satisfy the payroll) will not be an adequate solution. Equally, adequate turn-round of a failing company costs both a significant amount of time and money, and neither you nor the investor will want to admit that the business is failing. Thus the fateful day of admission may well be when things are quite bad, perhaps beyond salvage. At this point, a damage limitation exercise is in order, in the face of what will probably not be clear headed and decisive pressures in the environs. First, get to the hub of the problem. Do you have an adequate product portfolio? Are the products selling? If not, is it an inadequate sales force or weak demand? Are the managers and staff working to expectations? Do you give the business appropriate direction? Have your costs escalated? In any event, you should take a long, hard, cold look at the business and prepare a new business plan. Present proposals for salvation to the investors before they do it to you. Remember, at this point the investors will be interested in saving their investment.

9 Make your technology a commercial success

Where there is no vision, the people perish.
The Old Testament, Book of Proverbs 30:8

Where do you go from here?

In this book we have introduced the basic principles that relate to protecting your intellectual property and taking it to the market place. This has involved preparing yourself for negotiations, selecting partners and deciding upon the commercial direction to take, i.e. licensing, or directly via a start-up company. It is important that the reader realises the enormity of the tasks that face the budding entrepreneur in today's competitive environment, and both the government and the universities have a significant part to play in the evolution of these businesses. Consistent policies towards the protection of IP and the financial benefits that accrue therefrom, a more competitive attitude towards patent protection (cf. US policy) and an understanding of the criteria facing potential investors are all areas where improvements are possible. In addition, universities must realise the importance and necessity of professional project management principles in the technology transfer process. The days of the privileged minority nineteenth-century scientists beavering away in an isolated room in pursuit of their own greater glorification are long gone. The reality of twenty-first-century science is one of goal-orientated, cost-effective team based research and development. The enormity and complexity of many of the projects now facing us affords no other option if the UK is to assume a competitive position in biotechnology.

Managing the company

As we move towards the end of this century and into the next, it is apparent that the dynamics of the market place are changing rapidly and it is likely that only those that are responsive to new markets will be successful. Businesses are becoming more and more customer orientated, and a philosophy whereby your company is totally responsive to customer needs is one that is crucial to success. Such values are inherently difficult to improve when already set in the culture and it is recommended that you attempt to instill these from the very beginning. The other key area of change is in the perception of quality. It is now not necessarily so that the lowest price will win the contract. True, it has to be a competitive price, but customers are increasingly looking for value for money. Thus it is not just the product or service that is involved but also its quality, reliability, the after sales service, speed of response and the availability. Similarly, products and services are increasingly required to be tailored to specific customer needs, so many now include individual service attention and optional (add on) specialist features.

The competition for UK biotechnology businesses is also getting fiercer and besides the EU, there are the American and Japanese markets to contend with, especially as these overseas companies are now targeting domestic customers. The key to our future competitiveness is the translation of the body of research into the commercial arena, i.e. how to identify which of these new ideas have true commercial potential. In essence, the successful business is driven by market demand or by the creation of new markets where none have previously existed; this is rarely driven by a technological push. Many of the potential competitor companies will view the world market as their sphere of operations, especially in biotechnology. As such they market their products and services globally, and often have elaborate agency and distributor networks to facilitate this. If you are to make your business a success, it should be more than a disguised research organisation, rather you should ensure that the performance standards associated with world class companies are both achieved and adhered to.

Steering the company in the right direction

It is important to instill values from the outset, since it is extremely difficult and expensive to change attitudes on working practices once they have been established. This is particularly so in a new growing biotechnology company where change and instability make management of the culture particularly challenging. Thus the following issues must be turned into a commercial vision before you employ your first person: (a) understand your customers and suppliers (if any) and build a relationship with them, always looking for ways to add value to the service you are giving; (b) look at your competitors and those in related fields and learn from their successes and mistakes; (c) devise strategies to shorten the lead times towards new products and services; (d) decide upon your core business and concentrate resources upon it; and finally (e) make sure that all employees share the preceding values and vision. This should include attention to quality, improvements in products and services, participation in the derivation of new ways of working, a will to succeed and the desire to be part of the building of a significant business. All of these points are related to the strength of the company culture; the encapsulation of the inherent values and philosophies of the organisation.

Company culture

There are many definitions of company culture that generally relate to historical patterns of symbolism and inherited conceptions, by which means communication regarding the organisation is both potentiated and perpetuated. If you consider that the culture is something that an organisation *has*, then it is a variable that can be changed and influenced. If on the other hand you view culture as what the organisation *is*, then it is a continuously evolving myriad of subcultures, and as such, it is more difficult to change. In the initial stages of a biotechnology start up, the former is more likely to be true and fortunately, given the correct blend of characters and a sufficiently powerful vision, you are in a position to create a new culture. If you establish the common goals at an early stage, then you will have markedly improved productivity over those who do not.

Several management studies have indicated that there is a correlation between commercial success and a strong culture. Cultures can be classified (Handy 1985) as: (a) the 'power culture' whereby the central figure is the founder of the organisation and all power and decisions regarding resource allocation emanate from this point; (b) the 'role culture' where the power base lies in the position of the individual and bureaucracy is likely to predominate; c) the 'task culture' where the culture is project orientated and seeks to coordinate the resources of individuals with expert knowledge; and (d) the 'minimal structure' culture where individuals are relatively autonomous. In the start-up company the power culture will predominate at first. However, as the organisation grows, the pressures on the 'Zeus' figure will increase and unless delegation is practised effectively, the rate of progress will be slowed down. This is because Zeus very often views the company as a personal possession and not an entity into itself. As such he finds it difficult to trust subordinates and cannot let go of power. The net result is that his management becomes slower and more inefficient, and the employees find him unresponsive and aloof. However, this is symptomatic of a failure to pervade the business vision throughout the organisation. The power-culture phase should last for a short period of time only. The quicker you move to a task-orientated culture in an R & D based biotechnology start up, the better it will be for morale and productivity. In reality, this requires a leader with enough managerial confidence to pick the right people, to believe in them and to set and manage the culture. This is unlikely to be achieved by someone who is not wholly devoted to the building of the business, and many biotechnology failures are probably due to the prime mover being an academic who can only devote 5 to 10% of his/her time to the venture due to other commitments. The task culture requires project managers who can lead junior staff (i.e. not just assume a position by virtue of their name being in a box on a particular organisational chart), who behave in a manner consistent with the vision, who can promulgate these values through to the staff who and can maintain the vision when trouble shooting is required. Since these people must also have the necessary mix of technical skills to maintain academic and managerial credibility, they are extremely hard to find. Under each project manager you may find a subculture developing, e.g. competition or rivalry with another project team. This must be managed carefully, as it can easily turn sour and you must at all times retain a spirit of mutual cooperation. This is best achieved by regular inter-project meetings, in addition to intra-project meetings, by mixing up project personnel in the laboratories and by ensuring that all resources (consumables, capital equipment, offices, library and refreshments) are communal. Again, this must emanate from the instillation and reinforcent of the cultural values by the founder (Zeus) in weekly meetings with the lieutenants.

As the organisation achieves a considerable size, the task culture will tend to become a role culture, as is observed in many pharmaceutical companies. The size and complexity of the processes undertaken in such businesses seems to necessitate an increase in bureaucratisation in order to maintain communication, control structures and record keeping. It is perhaps ironic that many of these organisations are striving to return to a task-based culture, with a variable degree of success. Herein lies a fundamental reason for the paucity of high quality research in the pharmaceutical industry when compared to the biotechnology industry and an explanation for the changes that are currently occurring in the industry, with the pharmaceutical sector increasingly being development orientated and with the research being contracted out or carried out independently by a plethora of small biotechnology companies.

It is wise to avoid the minimal structure culture, since, whilst this has naturally been assumed by the academic sector, it is not consistent with the commercial goals of the biotechnology company. Indeed, it could be argued that this structure is inappropriate even for academic laboratories, the evolution still being driven by Victorian ideologies and not the competitive challenges of the here and now.

Creation of the culture

So how do we begin to create a new culture? It is important to realise that no single approach is likely to be effective in culture creation or management and a summary of some will be given here.

☛ Cultural heroes

The role of the cultural hero is often overlooked but subtly they provide: (a) role models; (b) a symbolism of the company to the outside, thus helping to attract the required personnel; (c) a method for cultural preservation, such as the legacy left by a historical figure; (d) an example for standards of performance; and finally (e) an example of how goals can be attained and credit correctly apportioned. The cultural hero plays a quite different role to that of the manager and can potentiate the culture from within, thus providing a key catalytic function. How then can heroes be selected and created?

First, select a particular characteristic such as a proven innovative ability, the ability to command respect or perhaps popularity. Second, accentuate these characteristics by giving the individual a high profile in the organisation. This could be achieved by publicising notable achievements, word of mouth story telling, and encouraging the individual to assume the role. Alternatively, you may like to identify and potentiate the achievements of existing heroes, such as the founder, or indeed, yourself.

☛ Rituals, ceremonies and social events

The role of ritual and ceremony is often deep rooted and is an important method for releasing tension. It is thus difficult to create such cultural aspects but one has to do so. The new management must be seen to set standards and be seen to maintain them, particularly attitudes towards language, interpersonal behaviour, presentation format and disruptive behaviour. Have a clear vision of how you see the people working in your company. It is suggested that you attempt a flexible format (first name terms), a professional dress rule and at all times encourage those who are helpful to their colleagues. Try not to ritualise coffee and tea breaks (where a lot of time is wasted and damaging gossip can be potentiated) and adopt a flexitime system, whereby employees are requested to be on site at certain core times such as between 10 a.m. and 4 p.m. This will facilitate the arrangement of meetings, etc. The autocratic clock-watching mentality is inappropriate. When you have achieved a critical mass of say greater than 20 people, you might consider the award of a plaque for outstanding achievement. This not only creates internal competition in a humorous way, but also reinforces the culture. Another beneficial technique is to target selected groups toward a particular trouble-shooting assignment. Chosen from across your employee spectrum to represent subtle cultural movements, you can effect cultural change by predisposing the task force to come to the intended end point using its own volition.

Social events (e.g. Christmas parties) have an important role in normalising the culture and allowing the release of accumulated tension. Discussion of particular characters allows the development of cohesion and sets of values amongst particular groups. Whilst one must not let this go too far, with the inevitable risk of polarisation of particular groups, it can be manipulated and directed so as to strengthen the culture. For a short period, the conventional stratified roles that are based upon position, salary, etc. are neutralised through social activities that will release latent intimacies. The positive atmosphere thus engendered may then appear in the everyday interactions, perhaps making senior figures more approachable and improving the routes of communication.

☛ Company values

The company values represent its philosophy for success and should be both known and shared by all employees. They are often neither written down nor definable in the same sense as procedures, policies and structures, but nevertheless represent a common feeling of pulling together. Core values often mean little outside the company but the internal effect can be dramatic; if you wish to strengthen the culture, do so via a suitable slogan to be permeated via notice boards, memos, word of mouth, etc.

☛ Communications

The expression of our cultural aims to the organisation is key to its implementation. The primary method is to utilise the endogenous network of storytellers, gossips and spies, e.g. the use of secretaries to communicate cultural events to the management. Ideally, before you effect cultural change you should have a network of information conduits in place since only by staying closely in touch with the culture can we attempt to make the correct decisions and judge success.

☛ Symbolism

Within any organisational culture there are a number of symbols that send signals to the individuals within the culture and also to those outside. For example, senior management with plush offices are symbolically trying to convince others of their position of power and control. As ostensible culture managers we must be aware of the effect of these symbols on others. For example, a junior employee may be intimidated by being called to a plush office for a meeting. The question is, would it be more effective to communicate the cultural message by holding the meeting in less threatening surroundings?

☛ Is a cultural change necessary?

As a manger of culture, it is essential to realise when you should react to environmental change by adjusting the culture. You must determine whether: (a) the change is likely to be maintained; and (b) if it necessitates a change in the company culture. To maintain the *status quo* in the face of change, you must potentiate and strengthen the existing culture using the aforementioned methods to maintain an air of consistency. What then are the signs of an ailing culture and when is direct action necessitated? These symptoms could include: (a) different cultural values in different parts of the company; (b) disruptive 'heroes'; (c) disorganised rituals, e.g. key personnel either not invited or not present; (d) an inward focus, e.g. attempting to score points off colleagues; (e) a focus on short-term goals; (f) a lowering of morale; and (g) emotional outbursts.

First, the attentive culture manager should consider openness, tolerance of failure and concern for the individuals well being, all of which removes potential foci for resistance to change. Second, it is important to involve the employees in the planning and implementation of change since this will: (a) reduce resistance to change; (b) enable a rapid return to maximum work rate; and (c) reduce absenteeism or staff turnover during and after the change period. This can be effected by using a cultural hero to lead the change process, making major structural changes obvious from the outset, offering security of employment and being patient, since different people adapt to change at different rates. As the maturity of your organisation increases, so will the problems of cultural change and eventually you will need to employ a consultant to handle the process for you. In the embryonic biotechnology company there is both a unique opportunity to manage a culture and to sow the seed for the eventual success or failure of the company. The founder may create a culture in their own image to reflect their own values, priorities and vision of the future. Management studies have shown that the personal values of the founder not only become shared by the employees but remain when the business is modified, grows or changes leadership. Thus the manager in such a culture would do well to focus attention potentiating the image of the leader figure, since therein lies the strength of the culture. Indeed, in the initial phases, the inertia that results from tradition, ritual and ceremony will not exist. Thus, it is imperative that you start the business off in the right direction although the founder must be of the correct personality disposition to formulate and carry the vision forward.

As the business develops from the entrepreneurial creative stage to the more formalised growth phase, so culture change must be effected, in order to focus on long-term goals and ensure long-term survival. The necessary changes often result in either the resignation, or significant dilution of the founder leaving a culture management transition of considerable properties. If the company has been successful, so the employees are likely to be content with the *status quo* of shared values. If unsuccessful, then the shared values will be less than adequate and the problem becomes one of changing an old culture or creating a new one. One could of course, sit back and let the culture naturally evolve. However, this brings with it the inevitable risk of business failure, since morale can easily be shattered by inept behaviour. Rather, it is recommended that you consider the proactive management of the culture, such that you create a vision in the work place, thus maintaining motivation and commitment from the employees. This is essential to business success.

Structure and management

The business plan will indicate the number of employees you expect to have in the company, together with the positions of senior management, for example, the managing director and heads of finance, research and development, and marketing. You may have even proposed an organisational chart delineating the positions of the more junior employees. This plan must now be implemented by introducing the structure into the workplace and turning it into a functional unit. It is not simply a matter of employing someone, informing them of the position and leaving them to it. You must instill your vision of the company and create the culture, not by imposition, but by mutual consent. To do this, one must employ people with the correct mental attitude for a biotechnology company. You will find that there are a number of individuals with reasonable academic/professional qualifications and relevant work experience but few with either the right disposition, blend of experiences or dynamic personality. Sifting the applicants is a long and somewhat difficult task, since very often all you will have access to is a curriculum vitae that looks very similar to all the rest. In addition, there is a tendency to match like with like and if indeed you originate from the world of academia there will be a tendency to select those that are most akin to your own experience. In general, this will be a mistake for the evolving biotechnology company and it is imperative to realise that other sets of skills are required. In this regard, it will be wise to retain the services of one skilled in building multidisciplinary teams.

Recruitment and selection

The first employees in a new company are probably the most crucial appointments you will ever make, since it is through these individuals that your culture will be created and the company will be built. It is recommended that you first consider appointing a laboratory manager with responsibility for ordering, negotiating with suppliers and organisation of the laboratory. New arrivals, especially those of postdoctoral level, are easily disorientated in a new environment and are apt to continually make comparisons to previous positions of employment, which, if left unchecked, can result rapidly in reduced motivation. It is the management of such professional specialists that will become your prime activity; it will be much easier if there is both structure and a modicum of capital equipment for them to begin work. This is not to suppose that there should necessarily be a fully functional laboratory; often participation in the choice of major items of equipment can engender commitment and a sense of identity in the future.

Somewhere within the company, it is important to have a competent personnel officer from the earliest stages. Again, this is crucial to the well being of future employees. In particular, this person should be responsible for administration of employment contracts (P45s, etc.), job descriptions, pensions and the like. In the initial stages this is likely to be the responsibility of one of the directors or the company secretary, since there will be little for this person to do until a critical mass of say, > 25 employees, is reached.

As the company grows so you will employ project leaders (probably experienced PhDs) with support staff such as postdoctorals, research assistants and the like. A common phase used in UK science is 'technical', which often has a rather derogatory tone associated with it, referring back to the days of UK science when the scientists wore white coats, technical staff wore brown coats and there were separate canteens and toilets for each. Fortunately, some of these barriers are breaking down, but we are still left with a dual career pathway, one for technical staff and one for managerial staff. Any such structure is fundamentally flawed in the biotechnology start-up environment since you must engender the complete commitment of all staff. The creation of artificial barriers via assumed hierarchy is an inappropriate anachronism that will lead to reduced efficiency. How then to solve this problem? First, create a series of grades with overlapping salary bands. A total of four grades should be sufficient in most start ups. Second, give all staff a similar name such as 'Research Scientist'. Obviously you should have a separate name for administrative staff, but it is important to ensure that the grades are the same. Within these grades you will have the accountancy, bookkeeping, marketing, secretarial and maintenance staff. Hence one might then have an experienced non-PhD worker on a higher grade than a newly qualified postdoctoral. This is, of course, quite appropriate if at the time the former is contributing more towards the development of the company than the latter. With such a structure all staff will see that their grade and hence salary is directly linked to their contribution, which will be a mixture of qualification, experience and effort expended, not simply assumed as a right associated with qualification. There are two major techniques for potentiating these ideals, those of job design and performance appraisal, both of which will now be considered.

☛ Job design

It is not sufficient to employ someone and give them a vague verbal notion of what is expected of them in the workplace. This will (and often does) lead to a lethargic, unmotivated employee, which will inevitably result in managerial criticism for poor performance. Neither is it sufficient to write down a reporting structure and a list of duties, the so-called job description. Indeed, the job description as such is a document that should be torn up and consigned to the archives, never again to see the light of day. If your culture is strong enough and hence the leadership astute enough, then you should fully expect the organisational momentum to carry the new person through the job description barrier. However, there are few organisations where enough attention is given to the culture and as such, the job description will be with us for sometime, more often than not as a security blanket for a 'defensive' employee. Instead, one could perhaps consider a nine point plan for job design. (1) The vision. Management must communicate a clear strategic vision of where the company will be in the future, its product range and a realistic projection of its chances of success. (2) The positive development programme. This relates back to the four-point grading system where there is a skill and contribution linked remuneration system, whereby employees would be compensated for learning new techniques, bringing in new ideas and extending their existing skills in areas where a realistic contribution to the business could be made. This reduces the possibility of job demarcation and territorialism whilst increasing the opportunity for team based working. (3) Open plan laboratories. Design the new laboratory to allow maximum interaction. Within the company mix different projects up in the same laboratories, ensure office spaces are open plan and do not allow for all personnel of a particular grade to be in one room. Make sure there are communal facilities, e.g. canteen, coffee room and library/meetings rooms for all employees. (4) High output team working. Establish autonomous groups for each separate company project, with the number of personnel deployed being dependent on its size. Try to ensure that each person is involved in more than one activity so that you can encourage cross fertilisation between projects. Allow each member of the group to police itself and hopefully, you will be able to rely upon the strength of the culture to correct disciplinary indiscretions. At all times employees should be encouraged to develop their own range of skills and in particular, to help others to do so also. (5) Adopt a non-authoritarian management style. Ensure that management focus their efforts upon encouragement of group autonomy. This requires a considerable degree of patience as groups develop their own managerial style, some faster than others. (6) Make sure there is an adequate support structure. If you provide adequate central services (networked personal computers, libraries, wash-up, media making, DNA and peptide synthesis) then all project groups will function more efficiently. If you like, this can be viewed as part of the 'taking away excuses' philosophy, whereby barriers to progress are removed by the management such that the focus is solely on the project. (7) The project management system. It is essential to have clearly defined goals, including milestones, and to review formally the progress made at regular intervals. A review every three months should prove adequate. Clearly define the milestones, not only in terms of the results expected from the allocated resource but also in their quality. Encourage feedback from the employees in terms of what they think they can achieve and then set the task slightly harder. In this way the team will be pushed to achieve, but if it does not do so the sense of failure will be low. It is most important that the milestones are suggested by the team members, however you care to engineer this, since whatever is imposed by management will be resented to some extent, even if the task is relatively simple. (8) Consultation. The employees should be consulted with regard to working practices from the earliest stages. As this policy matures, so a product champion will emerge engendering commitment and enthusiasm via a feeling of 'ownership'. This must be carefully managed so it does not become a possession, but again, the culture should be strong enough to prevent this occurring. (9) Skillmatching. Each and every project will have a unique set of skills that are required for its execution. It is the responsibility of the project manager to assemble the team with the correct balance of practical and personal skills. The latter will be discussed in the context of team roles later in this chapter, but with regard to the former the project manager should first assess the mix of skills necessary for successful completion of the project. Since all well-run organisations are run 'lean', i.e. resource limited, it is unlikely that the absolute ideal blend of skills will be available. Thus there will be a requirement for training in certain areas. If employees reach a particular standard as a result of this training and then contribute to a successful project, then they should be seen to be rewarded.

Thus overall, in the design of a new job, you should establish the vision of the company and make it obvious that there is: (a) free and open communication with the possibility of significant improvements in personal skills; (b) team and project based working with significant autonomy; (c) an attitude of 'if its not here, we'll get it' from the management; and (d) frequent consultation. Then you can begin to match the individuals' skills with relevant projects, find out how they would like to improve themselves and allocate group responsibilities if appropriate. If you get this right, then a job description will be unnecessary. These factors may be reinforced by a suitable performance appraisal system.

Performance appraisal

The performance appraisal system must reflect reward for adherence to the cultural values of the company. A performance appraisal form (Fig. 9.1) is designed to bisect the culture of the academic research laboratory and that of the highly structured pharmaceutical company. In essence there are elements of both that are required in the young biotechnology company, but either in its entirety would be inappropriate. As shown, the emphasis is placed upon co-operation rather than competition, and modified performance is reflected in the reward system usually by direct salary increases or decreases. The execution of the appraisal should be given considerable thought, particularly since any given manager is open to accusation of either bearing a grudge or favouritism. For this reason an appraisal is best carried out by an independent assessor but this has the limitation of the appraiser not being close enough to the appraisees work. In so far as is possible, try to ensure that only one person carries out the appraisal for the company (this is reasonable if the number of employees is less than 20) since this should ensure consistency.

Performance appraisal

Name:
Project:
Manager:
Period: From: To:

Criterion	Maximum Score	Actual score
Achievement of specified goals as determined by project milestones	45	
Demonstration of improved technical competence and flexibility towards new issues	15	
Role played in helping others to achieve their goals	15	
Ability to get on with others	10	
Personal qualities:		
Commitment	5	
Initiative	5	
Self discipline	5	
Total:	100	

Fig. 9.1. An example of a performance appraisal form.

Selecting the team

One of the most difficult areas associated with building a biotechnology business is the construction of the team and encouraging them to work together to best effect. It is not just a matter of selecting relevant skills, but of choosing the correct personality types for optimal functionality. Indeed, the latter is probably more important since the former can be acquired via a training programme. There are a series of eight roles that individuals naturally adopt when placed in a team based environment, as described by Belbin (1981). These describe the behaviour patterns between team members where the interactions affects the progress of the team as a whole and are as follows:

1 The company worker. A person who turns plans into practicality; systematic and efficient.
2 Completer. Such a person scans for mistakes and omissions, looks for work that needs a lot of attention and accentuates any sense of urgency.
3 Chairman. Essentially a co-ordination role, noting the strengths and weaknesses of each team member and helping each to achieve full potential for the general benefit.
4 Monitor. Such a person likes to analyse problems and to evaluate ideas and suggestions.
5 Resource investigator. This person likes to explore ideas and obtain information from outside the group, creating useful contacts as they do so.
6 Plant. This rather obscure name signifies the role of such people in the proposal of new ideas and concepts, often with attention to major strategic issues. Hence the concept that 'plants' give forth a multitude of blossom. Such people are often

highly creative and can often see solutions to difficult problems that are often opaque to others.

7 Shaper. These persons direct attention to the setting of objectives and the imposition of a structure onto group activities.
8 The team worker. Such individuals support other members of the team by building on their suggestions supporting weaknesses, helping communications and generally ensuring a good team atmosphere.

If you look at your colleagues around you, no doubt you will see some of these characteristics. The self-perception inventory in *Management Teams* by Meredith Belbin will provide an illuminating insight into your own personality and what you can best contribute to a team. From my own experience of this inventory it became obvious what my strengths and weaknesses were. Since that time, instead of trying to be all things to everyone (and not making a very good job of it), I have endeavoured to concentrate on my Belbin team strengths and, depending upon the circumstance, build teams which complemented the missing functions. In doing so, you will find that you do not recruit those in your own image, but those who supplement your own skills. In the efficient and effective functioning of a biotechnology team, one can easily see the necessity for influence of each of these personality types. However, the inappropriate construction of a team may have equally serious consequences. For example, a team with a highly creative ideas person (a plant) may easily fail or lead off in the wrong direction if these ideas are not attenutated (the monitor), if they cannot be reduced to practicality (the company worker) of if there is no structure to the project (shaper). Equally, excessive contribution from a completer may lead to a lack of forward vision (provided by a chairman) or a lack of creativity (provided by a plant). Therefore, beware of teams that are too heavily biased in one direction. Learn to see your employees not just in terms of their technical skills but also for their individual personality traits, which will dictate the efficiency of their performance within the team.

Leadership

One should distinguish the above personality profiles from another highly important consideration in the newly formed company; that of the position of leader. The latter can neither be given nor delegated. Inevitably one or more of your employees will come to the fore as a natural leader. The purpose of this section is to help you recognise the characteristics, nurture them for the best advantage of the company and perhaps even to help yourself assume the role to greater effect.

What then are the characteristics you should attempt to develop as a leader or what characteristics should you be looking for amongst your employees? To command respect of those around you it is important to have a high standard of personal and professional ethics (the two should be synonymous). Be totally uncompromising in upholding your standards of personal ethics and fight anything that suggests untoward behaviour. If you have made a genuine mistake, admit it. This will be respected by your colleagues and only the most insecure under-achievers will brand you incompetent. If you have won the hearts and minds of your staff, then they will be glad to see you are fallible too; it relaxes the pressure on them, making mistakes both less probable and more likely to be revealed. There are few attributes that are more damaging to the culture than having a mistake covered-up and not rectified through fear of reprisal.

The excellent leader will also have considerable common sense and an eye for reasonable behaviour. This means having a considerable degree of trust in your own judgement and being able to rise above a problem to see the whole picture in a broader context. Much time is lost in the workplace by petty behaviour and we have all had experience of such matters. Good leaders show a disdain for pettiness and get on with the job in hand, refusing to play the 'little games' that abound in most organisations. Towit, leaders are generally accomplished at prioritising both their own work and that of their colleagues. Furthermore, they tend to deal with the difficult problems first, believing that the shorter a painful experience and the quicker it is dealt with, so much the better.

Leaders are notoriously courageous, responsible, committed and dedicated, having little time to be depressed, being driven by the thrill of the job. Usually they are also creative individuals, tinged with an element of eccentricity making them obviously different from their colleagues. Those with leadership qualities express an innate, contagious enthusiasm for whatever they do but they are invariably level headed, are quick to learn and can bring order into chaos. Leaders are proactive rather than reactive, which again serves to aid the solution rather than compound the problem. The final characteristic that you should look for or seek to achieve is a genuine commitment to the development of the staff around you. The true leader supports the ideas of others and their educational aspirations, e.g. by allowing day release and paying for it.

As a leader, you should be the one who blazes the trail for others to follow whilst at the same time remaining fair and considerate of their feelings. Above all, be true to your vision, you may never get another opportunity quite like this one.

Employment and health and safety issues

When the financing has been completed you will need to action the business plan. The plan will have laid contingency for a certain number of staff that by now you will have chosen diligently. There are certain statutory obligations with regard to employing staff that will now be outlined. Employment law is another complex area and as before, if you are uncertain as to how to proceed, advice should be sought. As an employer, you have certain rights with regard to your employees, these being; (a) to expect honesty and no action to be taken against the interests of the company; (b) non disclosure of confidential company information; (c) a degree of competence, industrious activity and care of company property; and (d) that all discoveries whilst employed belong to the employer. The corollary of this is that the employer has a duty to behave reasonably towards employees, to have clear grievance and disciplinary procedures, to pay salaries and fees when it has been agreed to do so, to take reasonable care with regard to health and safety and never to discriminate with respect to sex, race, religion, etc.

Full-time workers are employed for greater than 16 hours per week whilst part-time is defined as being between 8 and 16 hours per week. Do not disregard part-time workers; there is a considerable pool of skilled technical staff available. For new employees you must provide an itemised statement with the pay that by law must include: (a) the amount of salary before deductions; (b) deductions for income tax (PAYE), national insurance (NI) and others as applicable, clearly identified as such; and (c) the amount of pay after deductions. The employer must act as collector of income tax and NI on behalf of the government, but otherwise cannot make deductions unless this is stated in the employment contract or requested in writing by the employee. Upon taking on an employee you must inform the tax office, obtain the employees P45 and work out the relevant deductions from the tables that will be provided. Within 14 days from each month end, the relevant amounts must be sent to the tax office and at the end of the tax year (5 April) a form will be sent for each employee that summarises the taxes for the year. This is the P60 and is due to be returned by 19 May. After one month of employment you are required to give minimum notice periods and pay guaranteed rates if there is no work, and within 13 weeks of employment there is a requirement for a written statement of the main terms and conditions of employment. However, it is not just this that constitutes the contract of employment; the latter includes any oral or written discourse, the contents of advertisements, items discussed at interview and any subsequent conversations regarding terms and conditions, the contents of the offer letter and the written contract itself. The written statement should include: (a) the name of the employer; (b) the employees name, the job title, the reporting route; (c) the date of first employment; (d) the rate of pay, interval of payment, hours of work (including those of normal working); (e) holidays (including public holidays); (f) conditions which relate to sickness and injury; (g) pension schemes; and (h) the period of notice that has to be given each way. In addition, there needs to be an indication of whether a contracting out certificate is available under the Social Securities Pension Act 1975, a written note of the disciplinary and grievance procedures if there are more than 20 employees and finally the name of a person to whom petition can be made if there is dissatisfation with the grievance procedure. After two years, an employee has the right to fair dismissal with written reasons and furthermore, there are particular provisions with regard to maternity. The Department of

Employment provides leaflets on all of the issues that you will need to consider.

There is an obligation to provide reasonable circumstances for health and safety provisions, both with regard to the premises and to the work itself. The first step is to inform the local Health and Safety Executive Area office of your business name and address. It would do no harm to communicate your business intentions and to seek their advice from the earliest juncture. Employer's liability insurance is essential and the relevant certificate must be displayed in the workplace. The workplace must be safe with regard to fire precautions, electrical fittings, first aid, etc.; these provisions are particularly important in the laboratory where there is generally a significant proportion of electrical equipment, solvents, dangerous chemicals and the like. It is essential to comply with the provisions of COSHH and a list of chemicals used must be maintained, together with a risk assesment, record of training for use, action to be taken in case of accidents, etc. These provisions include your obligation to install safety equipment such as fume hoods, goggles and protective clothing. If the company has more than four employees there must be a written statement concerning health and safety policy which should be on display, together with a poster from the Health and Safety Executive which delineates the relevant law. More than ten employees require that an accident book be maintained.

Bookkeeping and credit control

One function yet to be considered is that of bookkeeping. It is necessary to maintain a methodical system for financial housekeeping in order to help the accountant trace the business activity at the end of the year, to satisfy the tax authorities and for you to maintain close control on the business activity. There are four types of book that should be kept. (1) A cash book, for recording the income and expenditure, including columns for VAT. The difference between the two columns should equate to your bank balance and you should attempt to reconcile this at the end of each month. (2) A sales day book, to record who owes money to your business. Enter numbered invoices as they are sent and store them in a holding file. When paid, transfer them to a permanent file and record the event in the sales day book. (3) A purchase day book, to record the money that you owe. As before, record transactions immediately and file incoming invoices; then instigate a diary system to ensure that invoices are paid on time, i.e. according to the agreed credit period. (4) A petty cash book, for minor spending. This should record VAT and be reconciled with the cash book as necessary.

Once launched, the company must survive in a harsh world. Management of the cash position will now become the pre-eminent concern. The breakeven calculation is a useful tool, but the reality is that the goalposts will move as the business develops and so will the breakeven point. Even so, the business will only be 'ticking over' if it reaches breakeven and no more. The aim is to make a profit.

To judge your success a budget is required and this should be derived from your financial projections. At the end of each month, the actual figures should be compared with the budget; there are excellent packages available for the personal computer to enable you to do this. The comparison will enable identification of rights, wrongs, lessons for the future and potential problem areas.

A vitally important area for cash management in the business is that of credit control. Any period of credit costs money and although there may be an accepted practice in your area of business, the credit period should be reduced as much as possible. For example, this could be cash with order, cash on delivery, payment 30 days after delivery, etc. Encouragement of prompt payment can be via a discount for early payment or a levy on late payment. However, both have political consequences and such a strategy must be carefully considered. At all times you should ensure that invoices are sent out in a timely fashion and preferably with delivery of the services (e.g. the final report) or products. Furthermore, ensure that the credit terms are stated clearly on the invoice. Shrewd contract negotiation for long 'service' projects should result in interim payments with a balance payable at the end, perhaps according to results or some other pre-agreed target.

Occasionally you will meet 'bad payers', i.e. companies who do not settle their invoices on time. Many businesses liquidate due to inadequate credit control and this is an essential part of business management. If the volume of proposed business is large then it is advisable to do some due diligence on the customer before taking the order, e.g. ask for a bank reference, trade references (you should be able to find the level of financial risk that others will accept) or copies of the latest accounts. This information can then be used to determine a credit limit for each customer in much the same way as is common practice for use of storecards, credit cards, etc. Chasing bad payers is an unpleasant task, but unfortunately it is one that must be faced. First, politely request payment by facsimilie and at all times in writing. Second, if there has been no response within a week, then telephone the creditor to try to ascertain the problem, e.g. there may be a query on the account or the goods/services that you have provided. Third, call every three days for a few weeks, being aware of the use of 'avoidance tactics' and failing this, try to visit to collect the money in person. If this is impractical, then pass the matter to your solicitor and/or threaten to use a debt collection agency. The latter could have an adverse affect on the credit rating of the company and so may have the desired effect. Fourth, there is the option of actually using the debt collection agency or proceeding through the small claims court.

Your suppliers may occasionally wish to investigate you since they may also be cash hungry and unwilling to allow long credit periods. A new business may not be able to provide bank and trade references or the latest accounts and so expect to be required to pay cash on delivery, at least until you have built a trading record with the supplier. However, in molecular and cellular biology, there are likely to be many suppliers all competing for the business; terms are generally 30 days, which you should build into your financial projections.

Causes of failure

If ever there was a statement to tempt fate, 'causes of failure' is it. There are so many ways and reasons why a business might fail. Society as a whole does not look favourably upon such failures that may confer stigmas for some considerable time period. However, there is a lot to be learnt from such an experience and the management literature is full of 'rags to riches' stories, where by entrepreneurial individuals have failed in one venture only to rise successfully in another.

The points to watch for are as follows. (1) Underpricing the product. If the market is new you can try to skim the market with a premium pricing strategy. If there is competition in the marketplace, do not assume that you must offer a lower price. As an alternative, you may increase the price and offer a higher quality product. (2) Try not to overestimate sales and err on the prudent side at all times. Similarly, try not to underestimate how long it will take to achieve the predicted sales targets, it will probably take longer than you think. (3) Another common underestimation is that of costs and the concurrent failure to control them. It is all too easy to find you require a resource that you had not considered previously or that there are unexpected increases in certain costs. (4) A lack of skills in particular areas is also a cause of failure. Entrepreneurs all too often believe that they can cover all of the necessary areas (e.g. marketing, finance, production, technical) without help from others or cash constraints may serve to preclude such assistance in the early stages of business development. In any event, the entrepreneur should undertake some self-analysis and decide upon strengths and weaknesses. Confront the weaknesses and employ someone to provide the skills you do not have. (5) Probably inherent in the former point, a lack of marketing skills could lead to inadequate market research and consequent business failure. (6) Loss of cash control is also a key reason for business failure; holding too much stock, excessive credit periods and payment of debtors too promptly can all contribute. (7) Finally, businesses can fail by taking unnecessary risks, e.g. by diversification into areas with which you are unfamiliar.

Whilst the above points are not a panacea for success, they will perhaps help you to avoid some of the inevitable pitfalls.

10 Conclusion

The empires of the future are the empires of the mind.
Sir Winston Churchill
In *Onwards to Victory*, 1944

The biotechnology business environment

The biotechnology industry is now entering a renaissance, as the initial excesses of expectation evaporate and many start-up companies begin their journeys with more experienced management and better directed investment criteria. Hopefully this will lead to fewer mistakes in the market place. The science associated with biotechnology is burgeoning outwards at a considerable rate, and we can reasonably expect the number of licensing deals and the preponderance of start-up companies to increase. The change in UK government policy to both encourage and expedite the transfer of technology from our research base will help create a positive environment for biotechnology, thus allowing these opportunities to assume their rightful place as a significant contributor to national wealth and prosperity. If this text has given you an insight into these processes, removed some of the mystique of the commercial world and helped facilitate the transfer of your technology into the market place, then it will have succeeded in its purpose.

Perhaps more so than in any other industry, research is the lifeblood of biotechnology. The basic research breakthroughs of the late 1970s and early 1980s have opened an unprecedented Pandora's box of opportunity, procedures and applications that were once in the realm of fantasy, but are now with us. Ten years ago, who would have thought that we would be able to stimulate red blood cell production with molecules such as erythropoietin, thus giving the opportunity to reduce the need for blood transfusions? Or accurately amplify DNA from minute samples as in the polymerase chain reaction? The new technology has of course not been without its critics and a number of organisations have been incorporated in order to monitor technological progress. As academics and biotechnologists we all have a responsibility to portray our research in a socio-economically favourable manner and this particularly applies to dealing with the media. The criticisms raised by these bodies are often genuine concerns from people who will ultimately benefit from the technology and as such, must be addressed. For example, 'patent concern' has been set up to consider the issues of patenting of genetically modified organisms and has made statements on these issues. The Ha-*ras* transgenic mouse (Oncomouse) was 'designed to suffer' according to one report, which is indeed a valid emotional response. Similarly, 'parents for safe food' has raised concerns about the use of bovine somatotrophin. The short-term and local economic benefits of genetically engineered crops are perhaps obvious but the knock on effects are less so. We cannot (as scientists) predict the biological outcome of release of herbicide resistant and frost-resistant plant species such as corn or cotton. It is even harder to predict the economic outcomes from the commercial exploitation of less applied research. Genetic engineering may allow us to grow tea in Wales for example, which may help the Welsh economy in an unprecedented way, but would have equally unprecedented economic consequences elsewhere in the world.

Similarly, although perhaps a little more difficult to answer coherently, the World Council of Churches issued a statement in 1988 to the effect 'To claim a patent for a life form is a direct and total denial of God as creator, sustainer, breath of life, immanent spirit in the within of all being'. I suspect that many of us undertake our research in the interest of improving the lot of humanity and whilst there will be a spectrum of opinion as to whether it is justifiable or not, commercial exploitation brings forth a series of different issues and one must be very clear as to the benefits of the research in hand before you embark on a path to market.

The overall presentation of the commercial exploitation of biotechnology must involve:

1 Publicly available information and stringent controls upon release of such organisms into the environment.
2 Adequate public debate with adequate representation from all sectors of society.
3 Arguments relating to social and ethical issues in addition to those that justify medical or other need.
4 Designation of liability for mistakes and a policy on help for the developing countries via either funding or collaboration.

Anyone who is thinking of commercialising their technology must have a justifiable position on the above issues. Since commercial decisions are often based on public perception and the possibilities of market acceptance, you as the originator and maybe owner of the technology must be in a position to advise on these issues. Thus, it will be wise to have considered such matters beforehand. Indeed, the importance of public perception has been noted in a recent UK government report (ACOST, 1990 HMSO) as an acceptance of the potential encumbrance to commercial prospects of consumer resistance and fears regarding safety and pollution.

In the early days of the biotechnology industry there were concerns regarding the relationship between academic bastions and the 'sharks' of industry. In particular it was thought that the increase in biotechnology companies would result in a reduction in the rate of exchange of scientific information. This has not materialised, but for different reasons on either side of the Atlantic. In the US there is a grace period for patenting whereby a patent can be applied for up to one year after the work has been publicly disclosed. This suits both the academic and the commercial concerns who wish to see the technology protected for the purposes of either investment or further development. In Europe there is no such patent grace period although the slower uptake and lower awareness of commercial issues amongst the academic sector has not impeded information flow. Indeed, the 'publish and be damned' mentality has probably contributed to the less favourable climate for the biotechnology environment in Europe, since in comparison to the US there has been relatively little incentive for the investor who has been less able to protect the investment by building a base of intellectual property. From the latter part of the 1980s up to the present, the attitudes to and awareness of commercial issues amongst the UK academic community has improved, even if successful execution has been sparse.

A further early concern was that consolidation of new biotechnology companies by larger companies would lead to the worldwide dominance of a few large players over genetic products. This was seen as possibly detrimental to the needs and survival of Third World economies, many of whom depend upon non-genetically engineered products in a variety of ways. Such rationalisation has also failed to materialise and as the true extent of the new technologies are realised, so the industry retains its entrepreneurial start up and fragmented nature.

It is often said that the development of UK technology is our economic Achilles heel; whilst there is a strong (but reducing) science base, the translation of these discoveries into commercial products is generally poor. From the perspective of the inventor of a new technology there are two principal problems, both of which require professional help. First, there must be independent advice on how to finance the commercialisation of technology, whether this be as a 'start-up' company, a development project, or a licensing arrangement. Second, there needs to be advice on the workings of the technology transfer process and exactly how your research will be converted into a commercial reality. Relatively speaking, the involvement of the UK academic community with industry is lower compared to competitor nations such as Japan and the US. The relevant mechanisms exist, but do not appear to be greatly used, as there is generally a reluctance to seek the relevant legal protection (patents) that greatly aids the technology transfer process. Additionally, the relatively poor public perception of the benefits of biotechnology affect investment desisions when, all things being equal, other projects may be favoured over those within the biotechnology sector, where contentious issues may not be raised.

Thus to summarise, bringing biotechnology to the world's market place will require four major elements:

1 A strong scientific research base in each of the participating countries.
2 Mechanisms for transfer of the technology to a commercial laboratory capable of maximal exploitation.
3 A favourable public perception.
4 A more positive approach to the commercialisation of technology, both in terms of entrepreneurial activity and the regulatory environment.

During the 1980s there was an explosion in investment for biotechnology based ventures, particularly in the US. This was fuelled by high expectations for financial return, but the delay in providing profits that were in line with predictions has left a more sceptical investment community. Long-range forecasts of profit expectations are difficult in any business sector, but must be considered even more so in biotechnology, where new products are often introduced into untested markets.

The UK bioscience industry: successful flotations

There are three ways in which to improve the opportunities in the UK biotechnology industry: (a) to encourage the entrepreneurial spirit in our young and dynamic scientists, by helping them take the step into the world of commercial R & D; (b) by providing clear, efficient and credible mechanisms for technology transfer between universities, start-up companies and the grander commercial world, e.g. the pharmaceutical industry; and (c) by demanding more flexibility in the financial community regarding entry into the stock exchange. The latter provides the best possible exit route whereby the investor can realise that investment and given the prospect of such a route, there is a greater incentive to invest.

The UK has a number of features that make it attractive for a biotechnology renaissance: (a) the leading edge research community; (b) an established venture capital industry; and (c) a world-class pharmaceutical industry that will provide both funding and a source of corporate alliance. We have the opportunity to relive the US experience of the 1980s, learning from the mistakes that have been made and breathing in the successes. Indeed, despite the impression of volatility in the share prices of biotechnology companies, the returns to investors have been above the average overall. The problem for the investor is to pick the star performers, since there are several that have under performed. In any event, it is likely that the UK could provide a more rapid return on the investment due to the economies of the cost structures and the accelerated clinical trial processes. Indeed, there are likely to be flotations of UK biotechnology companies over the next few years, since several now have drugs in phase III clinical trials.

The management gap

It is often said that the critical shortage in commercialising technology is not in finding either the ideas or the finance, but in having the right management. When asked of the key weaknesses in biotechnology companies, most venture capital investors will focus on the management. A bad idea *could* work with good management. Even the best idea will not be successful given bad management. This applies right across the board, from the concept stage at the bench to full commercialisation. There are innumerable instances of poor management in the laboratory leading to inefficient use of resources and these are issues that must be addressed. It is as well to realise that no one person is likely to fulfill all of the necessary roles and a wide variety of expertises and experiences will be required, together with considerable team based skills.

The lack of a pool of good managers in this sector has several origins. First, it is extremely rare for academics to be trained in management. The whole career structure is oriented towards passing exams and obtaining PhDs, all of which encourages a competitive attitude towards ones colleagues. There is little or no emphasis on team based problem solving, with the result that many academic laboratories have a selection of students and postdoctorals all working on their 'own' projects often with a competitive attitude to those around them. This encourages, indeed selects, for a series of 'founder centred' power cultures that have little regard for teamwork. Those that are most successful in this environment are then given positions as laboratory heads, and expected to manage people. Thus not only is there no training in personnel issues, but the academic system has selected out those with the skills most applicable to managing groups of people. No wonder many find the transition difficult and end up with what would be considered to be extremely poor project management systems in other organisations.

Second, the rapid rise of the biotechnology industry has left a big gap in staff with experience of management functions such as sales, promotions and marketing, especially in small entrepreneurial companies. The number of such executives is increasing in the US, but the pool of suitable candidates is still low in Europe. Entrepreneurs and investors have not encouraged a significant number of those with managerial experience to move from larger companies in related sectors (e.g. pharmaceuticals) to the smaller companies. From their point of view, such individuals would be a ready source of at least some of the necessary skills. However, one must remember that the culture of a small biotechnology company is very different from that of a large pharmaceutical company and whilst many of the techniques, procedures and project management systems will be useful in awakening the academic wishing to enter this sector, they are not necessarily appropriate to the embryonic biotechnology company. Indeed, the hierarchical values installed may do more harm than good. In addition, the biotechnology environment is inherently more risky and those requiring a stable position (perhaps for family reasons) or those uncomfortable 'on the edge' may be dissuaded from entering the business or underperform when there. Other sources of skilled staff include those few who have acquired relevant training (e.g. in law or MBAs) or those who have, in addition to scientific expertise, learnt commercial skills on the job. The latter category have often felt their way along in difficult and uncertain circumstances. Unfortunately, those who both understand commerce and science are currently hard to find.

The financing environment, societal regulation and intellectual property

The questions we must ask and seek to redress are all concerned with how we in the UK can realise our undoubted scientific potential in the world's biotechnology marketplace. The basis of the US advantage lays in the historical sophistication of their financial markets. First, US equities are more liquid than those of the European counterparts and the culture reflects a greater willingness to take risks. Second, the US business community has a more positive attitude to the benefits of the services of professional advisors who are in touch with the needs and wants of the markets, who can help in arranging finance and most importantly can help discover sensible interactions and negotiate the contracts. This is facilitated by several biopartnering conferences held each year, whereby contacts can be made and relationships established. The European approach contrasts, in that professional advice is less likely to be sought until negotiations are at a relatively advanced stage. The more fluid and interactive environment in the US is one that Europe should attempt to emulate.

In the business community there is often a tendency to reject outside technology, which has been referred to as the 'not invented here' syndrome. That is, if it was not developed in-house then it must be regarded with suspicion. This also occurs in the academic world whereby harsh judgmental decisions are sometimes applied to ones colleagues, often on the basis of little or no information. In contrast, the internal suspicion of the (untrained) laboratory head often leads to rejection of work from their own employees, only to find it gain credibility when published from another laboratory. How can we overcome these barriers to entry for your technology? The answer lies in marketing. Marketing yourself, your technology and your laboratory in a professional manner. Your sales pitch should be regarded as important as the research, since unless your technology comes to the light of day in the right place, it could be lost. The blinkered vision of technology that comes from outside the commercial world reflects issues of ego and fear of competition. Many companies will only entertain purchase of the technology if they can purchase it outright; the pros and cons of such arrangements will be discussed. We are however, in a progressive period when government policy is changing to accommodate improved interactions with industry. Both the European Union and national governments are looking for practical applications of the research that they have supported, an important step in devolution of this research into society for the public good and in improving national competitiveness. It would be surprising if biotechnology and the products derived therefrom does not influence all of our lives in the next century, from new medicines and therapies to food. Several 'innovative research' granting agencies have been set up to encourage small businesses, so the help is there. However, the major advances will come by linkage of industry with the leading edge laboratories in the academic world. It is now a matter of changing the attitudes of both parties, ingrained by years of traditional intransigence, to encourage them to work together for both mutual benefit and the greater good.

The intellectual property that you have generated must be protected. The only reasonable method for achieving this is via a patent although as we shall see, this protection is not always absolute. The situation in biotechnology has been compromised by the relatively slow response to determination of new patent policy in this area. In retrospect this may not be so bad, in such a fluid and constantly changing environment, it is difficult to know the correct direction. Part of the problem of course, unlike with more conventional areas, is that the use of biotechnology has introduced many shades of grey and the otherwise distinct boundaries have blurred. In addition, innovations have occurred outside of existing precedents and new legislation has to be derived. The problem is compounded by the veritable explosion in competing patent applications in the biotechnology area and consequently, enforcement of such patents when granted gives its own particular set of problems. Legal battles are extremely expensive and can result in a make or break scenario, with the winners becoming exalted and the losers condemned to a falling share price and irrecoverable loss of reputation. Over the past few years the courts have been setting reasonable precedents and companies have become more rigorous in formulation of their claims. We may confidently expect that these processes will continue to improve and if you are an inventor, do not be deterred from entering the fray. Perhaps the most interesting point is, however, that with the introduction of these precedents investors can now see real practicality in the protection of their investment. A good patent granted goes a long way towards reducing risk.

The next set of issues to be resolved relate to regulation of the technology. The single government for a single market has helped the US forge ahead in this regard whilst with the myriad of directives from the European Union and general disharmony of standards between European member states, Europe lags well behind. From an investment point of view, the lack of a coordinated regulatory policy pushes up costs, since there are inevitably more complicated and expensive routes towards the effective marketing of biotechnology associated products. Assuming the US and European price structures are similar for a given product, higher costs mean lower profits in Europe and as such, the investment is less attractive. Furthermore, European companies must make decisions regarding their strategy for market entry. Is it better, for example, to enter directly the US market in the hope of greater profits but face the greater competition, or to enter European markets, deal with regulatory issues in each country and have to devise an individual marketing strategy for each? Tough but not insurmountable. What is lacking of course is the pool of managerial staff with the relevant experience.

The way forward

Hopefully this book has give you a better idea of how to take your technology to the market place and how to exploit it to best advantage. As with nearly any subject, it is impossible to encapsulate every circumstance or eventuality. However, at the very least, you should now have a perspective on the relevant parameters.

The protection of your intellectual property via patenting and its valuation has been considered. The latter is inexorably difficult and varies according to market conditions amongst other parameters. It is in this area that you require the assistance of professional business development managers who know the field, can access relevant market information, the prices that are being charged, royalties expected, etc. From this information you can make the key decision as to whether to licence the technology or go it alone via a start-up company. With regard to the former we have discussed the content of the licensing document and how to go about finding licensees. For the latter the basic issues regarding the formation of the company, your role as a director, the contents of the business plan, the types of finance available and how to raise it have been discussed. Finally, attention has been given to the management of the company and the structure of the biotechnology industry. These sections are intended to raise some of the issues that will be prevalent early during the life cycle of the company.

It is anticipated that the contents of this book will be a helpful guide towards the commercialisation of your technology and of bringing it to the marketplace. Hopefully, you will derive immense satisfaction from so doing.

APPENDIX

1 Business terminology

Listed below are some of the terms commonly used in the business world. Not all are included in this text, but those that are appear in bold at first mention.

Assignment Transfer of ownership from one to another.

Board Meaning the board of directors of a company. These are the individuals who control the company for the benefit of the shareholders.

Burn rate The rate at which a company is spending cash, usually expressed monthly.

Business plan The strategy for the company that contains all the relevant information for the financing and running of the business. All companies should have a business plan.

Cash flow The most important aspect of any small business is the cash flow. The money coming in and the money going out constitute the flow of cash that determines whether a business will survive.

Citation If the Patent Examiner or opposition party requires access to other documentation, then these become the cited document.

Claims These define either the invention or the scope of protection in a verbal form. Usually, these are appended to the technical description and form an integral part of the application.

Closing Signature of the legal documents that binds your company and transfers the cash from the investor

Collateral The assets you pledge for a loan made to your company.

Compounding The effect of adding the interest on an investment to the principal each month or year so that interest is earned on interest.

Conception This is relevant to the US patent system and constitutes the first documentation of an idea, even before it has been transferred to the laboratory bench. However, this information must either be introduced into the US or conceived there.

Continuation The US patent system allows temporal extensions to patent applications during which more effective responses to the objections of the Patent Office can be prepared. The total of the original application is resubmitted as a 'continuation' or 'file wrapper continuation'. If lodged before the first application is relinquished the continuation is entitled to the first priority date. This process can be repeated several times.

Continuation in part (CIP) Again unique to the US, a CIP contains new information over and above that revealed in the first application. The precendent for priority date holds as for continuation in the case of common subject matter, but the additional material has its own priority date. Since R & D is moving very fast in the biotechnology sector, this facility allows strengthening of the original application. Clearly though, during the subsequent time other developments in other countries or publication of relevant information may have occured and have a significant effect upon the CIP.

Control Owning 51% of the stock of a company or, from another perspective, owning enough stock in the company to control what management will do.

Covenant The paragraphs in the legal document state the things you agree you will do and those that you will not do.

Current ratio The ratio of current assets over current liabilities. Less than 1 to 1 usually means a problem. A healthy balance may be 2 : 1. Any more than this means that cash is inappropriately utilised.

Current return Income that is received monthly, quarterly or annually as interest or dividends as opposed to the capital gain portion received on an investment at the end of the investment period.

Default When you have done something you told your investor you would not do then you are in default.

Discount rate The interest rate used in present value calculations to convert future cash flows into today's money

Due diligence The process of investigating a business venture to determine its feasibility.

Equity This describes the preferred and common stock of a business. It is also used to describe the amount of ownership of one person or a venture capitalist in a business.

European Patent Convention (EPC) There are currently 17 states which have signed up to this convention that came into force in June 1978. These are Austria, Belgium, Denmark, France, Germany, Greece, Ireland, Holland, Italy, Liechtenstein, Luxembourg, Monaco, Portugal, Spain, Sweden, Switzerland and the UK. An application may be filed at any of the national patent offices and then, security permitting, it is transferred to branches of the European Patent Office in Holland and then Munich. When granted, a European patent becomes a collection of national patents, each of which are then regarded as individual items of intellectual property.

Grace period (patenting) This is a period before patent filing in which limited public disclosure can be made without prejudice to the grant of the patent. The disclosure must be traceable to the information to be contained in the application or to the applicant or legal owner of the information. Thus if someone else comes up with a similar idea and discloses it, then the grace period may not apply.

Grace period (investment) The period of time you have to correct a default.

Hurdle rate The return on investment that is necessary to compensate the investor for risk.

Internal rate of return (IRR) The discount rate that equates the present value of cash outflows with the present value of cash inflows.

Initial public offering (IPO) The initial offer and sale of a company's stock to the public.

Lead investor The investor who leads a group of investors into an investment. Usually one venture capitalist will be the lead investor when a group of venture capitalists invest in a single business.

Licence Authority to work/use items that are protected by intellectual property rights, typically in exchange for a consideration of value, i.e. without change in ownership.

Licensor The party with the authority (often the owner) to license the rights; the **licensee** is the recipient of said right.

Net present value (NPV) The discounted present value of an investment minus the required initial investment.

Official action This is the report on the application issed by the Patent Office Examiner. It will contain citations of prior art, together with possible objections to the claims and technical specification. The applicant must generally reply within a given time and if they do not do so, the application may be considered as forsaken.

Opposition In Europe, formal oppostion to a patent must be received (filed) by the Patent Office within nine months of grant of the patent. The US and Canada do not have such a process in their patent system and in Japan opposition may be filed before the patent is granted.

Paris Convention or Paris Union This consists of 127 states (as at 11 October 1994) that have ratified an 1883 convention 'The International Convention for the Protection of Industrial Property'. The most important point here is the international acceptance of the filing/priority date in any member state and the ability to file in any other member state within one year, and still be able to retain the first priority date. The articles also provide for equal treatment regarding the IP rights of foreign nationals.

Patent acceptance This occurs when official examination is complete. Depending on the country, the application may then either be granted or opened to third parties.

Patent application or specification The filing of formal documents to a patent granting authority, consisting of a technical description and a list of claims. The former must be understandable by someone of 'average' skill in the subject concerned. For example, a patent for a new vector should be understandable by the majority of molecular biologists.

Payback period Measures the number of years required to recover the initial cash investment.

Present value The discounted value of a series of future cash flows so as to account for the time value of money. Alternatively, the value of a future series of cash flows stated in terms of todays money.

Product protection This refers to protection of any substance or biological entity and is not related to the way in which it was prepared.

Product by process (PBP) This refers to the protection of the method by which a substance or biological entity is produced.

Reduction to practice This term has a legal relevance in US patent law referring to the transition from a 'patentable' concept to a prototype product or process. This transition must occur in the US for the invention date to be ratified therein.

Refile Should a patent be dropped for any reason, it may be refuted with additional information, but will receive a new priority date.

Renewal fees These are payable on a per annum basis for maintenance of a patent.

Return on investment (ROI) The internal rate of return on an investment.

Second- and third-round financing The later rounds of expansion financings that follow the start-up round of financings.

Sensitivity analysis Analysis to determine how sensitive is the return on investment to variances from the expected values of key variables such as sales, margins, interest rates, etc.

Strasbourg Convention The most important article to arise from this convention is the use of the phrase 'state of the art', which is an ever-changing concept intended to keep pace with developments and judge how novel and inventive the application is at any given point in time.

State of the art In a sense this provides a definition of patented, pending patents and unpatentable prior art, i.e. the benchmark by which new inventions will be judged. It refers to the sum total of all knowledge that is in the public domain before the patent is filed.

Syndication The process whereby a group of venture capitalist will each put in a portion of the amount of money that is required to finance a small business.

Upside The amount of money that one can make by investing in a certain deal is called the upside potential (more of a US term).

Warranties These are items concerning your company that you have told the investor are true.

2 Useful addresses*

Company information and start up

- ***For information on companies registered in England and Wales:***

Companies House,
Postal Search Section,
PO Box 709,
Crown Way,
Cardiff CF4 3TF
☎ Tel.: 01222 380801

To obtain a shuttle form for company returns Tel.: 01222 380045 or 380145

- ***For information on companies registered in Scotland:***

Companies House,
Postal Search Section,
100–102 George Street,
Edinburgh EH2 3DJ
☎ Tel.: 0131 225 5774

- ***For information on companies registered in Northern Ireland:***

Registry of Companies and Friendly Societies,
IDB House,
64 Chichester Street,
Belfast BT1 4JX
☎ Tel.: 01232 234488

* These telephone and fax numbers are to the best of our knowledge correct as at 16 April 1995.

• ***Company start up; any accountant or solicitor or:***

Jordan and Sons,
Jordan House,
Brunswick Place,
London N1 6EE
☎ Tel.: 0171 253 3030

The Company Store,
26 North John Street,
Liverpool L2 9RU
☎ Tel.: 0151 258 1258 or 0800 262662 Fax: 0151 236 0653

Department of Trade and Industry

• ***Regional Offices:***

North East,
Stanegate House,
2 Groat Market,
Newcastle NE1 1YN
☎ Tel.: 0191 232 4722 Fax: 0191 232 6742

North West,
Sunley Tower,
Piccadilly Plaza,
Manchester, M1 4BA
☎ Tel.: 0161 838 5000 Fax: 0161 228 3740

Yorkshire and Humberside,
25 Queen Street,
Leeds LS1 2TW
☎ Tel.: 0113 244 3171 Fax: 0113 233 8301

East Midlands,
Severns House,
20 Middle Pavement,
Nottingham NG1 7DW
☎ Tel.: 0115 250 6181 Fax: 0115 258 7074

West Midlands,
77 Paradise Circus,
Queensway,
Birmingham B1 2DT
☎ Tel.: 0121 212 5000 Fax: 0121 212 1010

South West,
The Pithay,
Bristol BS1 2PB
☎ Tel.: 0117 222 1891 Fax: 0117 229 9494

South East,
Bridge Place,
88/89 Eccleston Square,
London SW1V 1PT
☎ Tel.: 0171 215 0574 Fax: 0171 215 0875

East,
Westbook Centre,
Milton Road,
Cambridge CB4 1YG
☎ Tel.: 01223 461939 Fax: 01223 461941

Scottish Trade International,
Franborough House,
120 Bothwell Street,
Glasgow G2 7JP
☎ Tel.: 0141 228 2869 Fax: 0141 221 3712

Welsh Office Industry Department,
Cathays Park,
Cardiff, CF1 3NQ.
☎ Tel.: 01222 825111 Fax: 01222 823088.

Industrial Development Board,
IDB House,
64 Chichester Street,
Belfast BT1 4JX
☎ Tel.: 01232 233233 Fax: 01232 231328

- ***Eureka enquiry point:***

The LINK Secretariat,
Office of Science and Technology,
Room 171,
Queens Anne's Chambers,
28 Broadway,
London SW1H 9JS.
☎ Tel.: 0171 210 0556 Fax: 0171 210 0557

- ***Additional useful numbers:***

☎ DTI Innovation enquiry telephone line: 0800 442001

☎ DTI Business in Europe Hotline (information on the single market): 01272 444888

Training and Enterprise Councils (TECs), business advice, consultancy, training. Contact local branches via the Yellow Pages or
☎ Tel.: 01742 701356, Fax: 01742 597 478.

Association of British Chambers of Commerce,
9 Tufton Street,
London SW1P 3QB
☎ Tel.: 0171 222 1555 Fax: 0171 799 2202
The above will provide help in finding contacts both nationally and internationally.

EC contacts

Guide to EC research funding
☎ DG XII Fax: 0032 2 2958220

RTD Helpdesk
DGXIII D/2
Commision of the European Union,
L–2920, Luxembourg
☎ Tel.: 00352 4301 33161 Fax: 00352 4301 32084

Financing

The British Venture Capital Association,
3 Catherine Place,
London SW1E 6DX
☎ Tel.: 0171 233 5212 Fax: 0171 931 0563

Venture Capital Report,
Boston Road,
Henley,
Oxon RG9 1DY
☎ Tel.: 01491 579999 Fax: 01491 579825

The Capital Exchange,
Wyvern Centre,
Barrs Court Road,
Hereford HR1 1EG
☎ Tel.: 01432 342484

Local Investment Networking Company (LINC)
4 Snow Hill,
London EC1A 2BS
☎ Tel.: 0171 236 3000 Fax: 0171 329 0226

LINC Scotland
30 George Square,
Glasgow G2 1EQ
☎ Tel.: 0141 221 3321 Fax: 0141 221 3244

Techinvest,
PO Box 37,
Dalton Way,
Middlewich,
Cheshire CW10 0HU
☎ Tel.: 01606 737009 Fax: 01606 737022

Eurotech capital. This scheme is adminstered by a network of venture capital companies who provide support for projects that arise out of RTD programmes. For information:
European Commission,
DG XVIII–J.M. Magnette, J. Berger,
Rue Alcide de Gasperi,
L–2920, Luxembourg.
☎ Tel.: 00352 4301 36261/46 Fax: 00352 4301 36322.

EC seed capital. A series of venture funds provide seed capital to promote the survival of new businesses, using evaluation criteria determined by the commission. The list of participants can be obtained from:
European Commission,
DG XXIII– Mr Richards,
Rue d'Arlon 80,
b–1049, Brussels
☎ Tel.: 0032 2 2960940 Fax: 0032 2 2966278

Association of Invoice Factors,
Northern Bank House,
109–113 Royal Avenue,
Belfast BT1 1FF

Association of British Factors,
147 Fleet Street,
London EC4A 2BU

The London Discount Market Association,
39 Cornhill,
London EC3V 5NU

Intellectual property

The Patent Office,
Cardiff Road,

Newport,
Gwent NP9 1RH
☎ Tel.: 01633 814000

and

The Patent Office,
State House,
66–77 High Holborn,
London WC1R 4TP

Derwent Information Ltd,
14 Great Queen Street,
London WC2B 5DF
☎ Tel.: 0171 344 2800 Fax: 0171 344 2871

Micropatent,
Cambridge Place,
Cambridge CB2 1NR
☎ Tel.: 01223 311479 Fax: 01223 301278

Chartered Institute of Patent Agents
Staple Inn Buildings,
High Holborn,
London, WC1V 7PZ.
☎ Tel.: 0171 405 9450 Fax: 0171 430 0471

For advice on biotechnology patenting:
Dr Julie Fyles,
William Jones Patent and Trademark Agents,
54 Blossom Street,
York YO2 2AP
☎ Tel.: 01904 610586 Fax: 01904 610909

Intellectual property insurance

Willis Corroon London Ltd,
Friars Street,
Ipswich,
Suffolk IP1 1TA
☎ Tel.: 01473 223000 Fax: 01473 222667

Intellectual Property Insurance Bureau Ltd,
Abchurch Chambers,
24 St Peter's Road,
Bournemouth BH1 2LN

Societies

The BioIndustry Association,
1 Queen Anne's Gate,
London SW1H 9BT.
☎ Tel.: 071 322 2809
Publishes the *UK Biotechnology Handbook* annually

The Institute of Directors,
116 Pall Mall,
London SW1Y 5ED
☎ Tel.: 0171 839 1233 Fax: 0171 930 1949

The Licensing Executives Society,
c/o Battelle Institute,
15 Hanover Square,
London W1R 9AJ
☎ Tel.: 0171 493 0184 Fax: 0171 629 9705

3 Sources of marketing information

There are a number of publications from which you can both obtain marketing information and market your own products, some of which are free.

The Best Database, Longman Cartermill Ltd, Technology Centre, St Andrews, Fife KY16 9EA.
☎ Tel.: 01334 477660. Fax: 01332 477180.

This consists of a database of active UK and European scientists, of technical services, of facilities available in universities, contract research organisations and companies and a CD–ROM database of current research projects. There is also a North American version of the database, a Eurotechnology newsletter and an Innovation journal, to which you may add new technology. The latter is for universities, medical schools and government research establishments to publicise new developments, free of charge. Access to the information is via a subscription service and is searched by commercial concerns who are searching for new ideas.

BioCommerce Data Ltd, 95 High Street, Slough, Berkshire SL1 1DH.
☎ Tel.: 01753 511777. Fax: 01753 512239.

Provides business information to the biotechnology industry such as financial abstracts, business news, biotechnology publications, mailing lists, on-line facilities and the UK biotechnology handbook, which is published in conjunction with the UK BioIndustry Association.

The British Library Medical Information Centre, Boston Spa, Wetherby, Yorkshire LS23 7BQ.
☎ Tel.: 01937 546364. Fax: 01937 546458.

Community Research and Development Information Service (CORDIS), which is accessed via the ECHO host. ECHO/Postbox 2379, L–1023, Luxembourg.
☎ Tel.: 00352 3498120. Fax: 00352 34981234.

• These telephone and fax numbers are to the best of our knowledge correct as at 16 April 1995.

This database contains information on EC RTD programmes. You can either add your results to the database in a search for a commercail partner in the Community or use the database to access the results of others. This can be especially useful if you are looking for technology to support existing programmes. To subscribe to CORDIS contact the RTD helpdesk.
☎ Tel.: 00352 4301 33161. Fax: 00352 4301 32084.

Coombs, J. and Alston, Y. R. *The Biotechnology Directory*, Macmillan, 1995. A directory of biotechnology companies, classified by product area and geography. Updated yearly.
ISBN 0 333 622049.

Derwent Information Ltd, 14 Great Queen Street, London WC2B 5DF.
☎ Tel.: 0171 344 2800. Fax: 0171 344 2871. E-mail: helpdesk@derwent.co.uk.
Access to the Derwent databases will allow searching in several areas, notably the world patents index, biotechnology abstracts, DNA sequence libraries, etc. Fees are payable for this service that has a wide clientele, such as patent agents, pharmaceutical and biotechnology companies.

The Discovery Database, Connect Pharma Ltd, Oxford Science Park, Oxford OX4 4GA. Tel.: 01865 784177. Fax: 01865 784178.
This is a free service to academic institutes whereby licensing opportunities in the pharmaceutical sector are brought to the attention of licensing executives in the industry.

Frost and Sullivan, 4 Grosvenor Gardens, London SW1W ODH.
☎ Tel.: 0171 730 3438. Fax: 0171 730 3343.
Online/CD–ROM information, pre-prepared report and customised research services are available in a numbers of disciplines, including pharmaceuticals, biotechnology and healthcare.

Guide to Biotechnology Companies. Published by *Genetic Engineering News*, 1651 Third Avenue, New York, NY 10128.
☎ Tel.: 001 212 289 2300.
Contains information on biotechnology companies, instrumentation manufacturers, law firms that specialise in biotechnology, venture capital companies, biotechnology recruiters and biotechnology consultants. Updated yearly.

Noticeboard. Published by the Oakland Consultancy and Publishing Services Ltd, 10 Jesus Lane, Cambridge CB5 8BA.
☎ Tel.: 01223 300475.
Published biannually and contains information from Unversities and research Institutes on commercial opportunities and centres of expertise in a variety of fields.

The Technology Exchange Ltd, Wrest Park, Silsoe, Bedford MK45 4HS.
☎ Tel.: 01525 860333. Fax: 01525 860664.

The exchange links sources of technology in a wide range of fields with those searching for it, helping to locate both licensees, licensors and manufacturers on a worldwide basis.

TRN News, European Association for the Transfer of Technologies, Innovation and Industrial Information (TII), 3 Rue des Capucins, L1313, Luxembourg.
☎ Tel: 00352 463035. Fax: 00352 462185.

Partially supported by the SPRINT programme, the database provides access to technologies in several fields.

4 Suggested reading

- **Baillie, I. C.** *Practical Business Management of Intellectual Property*, Longman, Harlow, Essex, 1986; ISBN 0 85121 131 3
- **Baillie, I. C.** *Licensing: A Practical Guide for the Businessman*, Longman, Harlow, Essex, 1987; ISBN 0 85121 132 1
- **Bainbridge, D.** *Intellectual Property*, Pitman Press, London, 1992; ISBN 0 273 03426X
- **Belbin, M.** *Management Teams – Why They Succeed or Fail*, Butterworth–Heinemann Ltd, Oxford, 1981; ISBN 0 7506 0253 8
- **Cary, L.** *The Venture Capital Report Guide to Venture Capital in Europe*, Management Today, Oxford (updated yearly); ISBN 0 9508734 5 4
- **Christie, A. and Gare, S.** *Statutes on Intellectual Property*, Blackstones Press Ltd, London, 1992; ISBN 185431 186 7
- **Gladstone, D.** *Venture Capital Investing*, Prentice Hall, Englewood Cliffs, New Jersey, 1988; ISBN 0-13-941428-2
- **Impey, D. and Montague, N.** *Running a Limited Company*, Jordans Publishing Ltd, Bristol, 1994; ISBN 0 85308 230 8
- **Handy, C.** *Understanding Organisations* Penguin, London, 1976; ISBN 0 14 009110 6
- *Realising Our Potential. A Strategy for Science, Engineering and Technology*. UK Government white paper, HMSO, London; ISBN 0 10 122502 4
- **Pearson, H. and Miller, C.** *Commercial Exploitation of Intellectual Property* Blackstone Press Ltd, London, 1990; ISBN 1 85431 044 5
- **Smith, G. and Parr, R.** *Valuation of Intellectual Property and Intangible Assets*, Wiley, Chichester, 1989; ISBN 0 471 61200 6
- **Spencer, V. and Binding, K.** *UK Biotech '94 – the Way Ahead* The BioIndustry Association and Arthur Anderson and Co., London, 1994.

Index

Major headings in bold

Printed in the United Kingdom
by Lightning Source UK Ltd.
114443UKS00002B/115-117

9 780521 466165